Leo's Lamb

by

Alexandria May Ausman

This book is a work of fiction. Any references to historical events, real people, or real places, are used fictitiously. Other names, characters, places, and events are products of the author's imagination, and any resemblance to actual events or persons, living or dead, is entirely coincidental.

Copyright © 2023 by Alexandria May Ausman

All rights reserved, including the right to reproduce this book or portions thereof in any form whatsoever.

Book cover design by Alexandria May Ausman
Editor: Jon M. Ausman

Library of Congress Control Number: 2023923248

ISBN: 979-8-9890048-9-8 (ebook)
ISBN: 979-8-9890048-8-1 (paperback)

Published By:

Ausman & Cousins, LLC
1700 North Monroe Street
Suite 11, Box 284
Tallahassee, Florida 32303-0501

For author interviews: ausman@embarqmail.com

The Das Kaiser Haus Series

Rise of the Priceless (Chapters 1 to 10)
Metal Illness (Chapters 11 to 19)
Jonas the Vampire (Chapters 20 to 29)
Prince of the Elders (Chapters 30 to 40)
Leo's Lamb (Chapters 41 to 50)
Mastermind Malfred (coming soon)

The Psycho Series

Cemetery Kid (Chapters 1 to 20)
Stop Calling Me Psycho (Chapters 21 to 33)
Motor-Psycho (Chapters 34 to 44)
Delusion of the Collar and the Key (Chapters 45 to 53)
Brutality's Prisoner (Chapters 54 to 64)
Aesthetic Akathisia (Chapters 65 to 74)
Metallic Burden (Chapters 75 to 83)

27 Masters Series

Anita the Benevolent (Chapters 1 to 7)
The Beast and the Witch (Chapters 8 to 16)
High Priestess of Schizophrenia (coming soon)

Book 5 Characters: Leo's Lamb

Abelard: a submissive of the Haus
Agnete: mother of Christian
Annette: a Haus trainee, Christian's first female love
Barnum: an Elder of the Haus
Beatrice: an Haus FemDom, friend of Julius
Ben: a Haus trainee
Bladrick: an Elder of the Haus
Casper the Ghost: an anonymous rapist
Christian: the anger and lust shard
Christian Axel: a Haus submissive, the Priceless
Claudia: a first floor Haus FemDom
Claus: an Elder of the Haus
Cora: a FemDom of the Haus
Debbie: Meine Liebe's sexual psychopathic and sadistic mother
Der Goldene Hund: the Voice or the Boss shard; the Conscious shard
Der Makellos: German Shepherd named "The Unblemished"
Dominic: a first floor Haus Dominant
Edgar: a black collar boy
Evelynn: a kitchen black collar
Felix: a black collar door guard hired by Peter
Gastov: a Haus Dominant
Geraldine: a Haus trainee
Gerard: stepfather of Christian Axel
Grisham: a Haus Dominant
Gretta: a Haus Elder, the Silk Queen

Gustov: a Haus Dominant
Heidi: A Dungeon Mistress; sister of Helga
Helga: a Dungeon Mistress; sister of Heidi
Hemmel: a Supreme Dungeon Master
Jonas: an Elder of the Haus
Julius: a Haus Dominant
Leo: a Haus Dominant
Louis: a first floor Haus Dominant
Mad Max: the sadistic shard of Maximillian, aka the Heart and Judgment
Mad Maxx: husband of Meine Liebe; a Haus Dominant
Mad Maxx: the masochistic shard, aka the Brain and Guilt
Malfred: a Haus Dominant
Max: the Soul shard
Maximillian: the seductive shard, aka the Libido
Maxximillian: the submissive adopted by the Elders
Meine Liebe: submissive and spouse of Mad Maxx
Milo: silver collar Haus seduction trainer
Nelda: a first floor Haus FemDom
Olaf: Haus black collar door guard
Peter: a Dominant of Der Kaiser Haus; best trainer of submissives
Rudolf: a black collar that works in the barn
Ryker: a deceased Haus trainee
Sofie: schizophrenic sister of Leo
Stefan: a Haus Dominant
Tamina: Malfred's valuable silver
The Lambs: Abelard, Annette, Geraldine, Milo, Ryker
Vilber: Haus black collar door guard
Xavier: the former Fur King of the Haus

Prologue

In book five of Das Kaiser Haus, Mad Maxx becomes the victim of repeated assaults from an unknown aggressor. He examines his long list of enemies trying to determine the identity of his stalker, while trying to discover the reason he has been targeted. As the attacks become more violent, Mad Maxx comes to the terrifying conclusion that his antagonist is one of the powerful leaders of the House.

Meanwhile, the Max shards and Christian continue Der Hund's mission by eliminating those that are standing in Mad Maxx's way. Extraordinary kindness from a once hated opponent opens Mad Max, and Max's heart to the possibility of a taboo love affair. The attention lavished on the Priceless by this most unexpected ally splits the Christian and Max shards loyalties toward each other. This war within Mad Maxx's head threatens to expose his secret diagnosis to those that would use it to prevent his escaping the House.

Jealousy, passion, brutality, fear, and madness run amok on the Elder's sixth floor fortress. Mad Maxx must carefully choose his alliances while finding his foes quickly both from within and from without. If he makes one mistake, he will be trapped in his metal as the Priceless collar for the rest of his life.

Chapter 41: Theft of Services

"I struggled with all my might as I was hauled backward to the silent, dark corner behind the back stair well. These steps were not heavily traveled like the main halls were as I told you. No one was even around to witness this foul abduction. I looked about wildly hoping that at least a single pair of eyes would see this horror and yell for assistance for me.

The hand clutched my mouth tight keeping all my sounds muffled. No matter how loud I tried to scream the hand kept my noises under the radar of ears if there had been any to hear my pleas. My assailant was not only strong, but they were also a big bastard to boot. I could tell I was being towered over and being held at little more than waist level of someone with much girth. I thought for sure this had to be a male, perhaps the brutes Vilber or Olaf?

I couldn't break free of, their grasp no matter what I did. Within only moments I was pulled out of the sight of anyone that would wander through that area. The unknown person threw me into the wall with so much force I smacked my head. That stunned me for a second, but I turned around quickly to attempt to get a look at the identity of this motherfucker.

It was dark in that place and my sight had not adjusted fast enough. I could only make out the shadow of a person of well over six-foot tall, broad, and coming at me rapidly. I let out a yelp of terror and tried to break and run away.

To my surprise the attacker sprayed an aerosol into my face with speed. My eyes burned with the fires of hell, and I suddenly could barely breathe. I covered my face with quick reflexes and fell to my knees choking and coughing, unable to even cry out for help.

The shadow came at me grabbing my upper arms and pushing me to my face. I thought for sure I was dying of whatever the hell he had sprayed into my eyes and mouth. I gasped and gurgled fighting with all I had to breathe while this bastard pulled my breeches down. I realized this sonofabitch was planning to rape me.

I tried to get up, but he grabbed the back of my neck, pinning me to the carpet. I clawed at the ground trying to find something to give me leverage to pull away. The brute forced himself into my chemically burned hind side without any kindness nor mercy other than a bit of saliva. I heard him spitting just before he began his sexual assault.

That pain of his harsh forced intercourse sent me into spasms worse than from his pepper spray. At that point I was burning everywhere. I couldn't even see who was doing this to me. I was blind with my eyes tearing up from the chemical and from the extreme agony I was in. My nose was stuffed up and my throat felt swollen shut. I could do nothing but gasp for air, cough, gag, tremble, and grab at the floor as he held me down with his heavy weight taking his pleasure with me.

It seemed like forever, but likely it was under seven minutes when I heard him moan out in his climax. I was still in pure hell, nearly paralyzed from the torture I was in. When he finished his brutal assault, he uncoupled then pulled up my breeches. He rolled me to my back and leaned over me. I tried to clear my vision blinking wildly to see the face of this monster, but I couldn't make out the color of his hair.

He reared back and back handed me. I put up my hands trying to protect my head. He grabbed my wrists and restrained my attempts to prevent his beating. I was knocked nearly unconscious before he stopped hitting me. I give up my fight expecting this was the end of Mad Maxx. I went limp and kept my eyes closed, seizing up with each breath.

This bastard decided that I was subdued enough for him to make his clean get away. I groaned out, nearly broken in two as the shadow slipped out of the corner, leaving me there to recover or die. I wasn't sure if he was still hiding in the darkness for many minutes. I hazarded that I was alone and rolled over to my face.

My stomach rolled then I vomited onto the floor with force enough to make me think I lost my guts there. I emptied all I had from the dinner the night before, then dry heaved awhile sure I was smothering. My breathing was getting a little easier, but my eyes still stung without any vision of worth.

I crawled with much effort to the wall and leaned my face into it while rubbing my eyes. I admit it, I cried like the lost kid from the trauma of that horrible scene. I tried to calm myself by figuring out who I was killing if I lived through this.

The shadow was too tall to be Vilber or Peter. The brutes girth was too slim to be that of Olaf. The man had moaned but I didn't recognize the voice of him. Other than that, I had nothing to go on. I wondered if Peter or Agnette had hired some black collar to teach me a lesson.

That didn't make sense though. If that was the reason for this rape and beating, then surely my attacker would have made such a thing clear. I was left terrorized, unable to determine who or why this had happened. All I could be sure of was that someone had planned, stalked, then enacted a brutal sexual assault. I shook in fear as I realized this man could do such a thing again unless I figured out his identity.

Within about thirty minutes of that attack my vision returned to me. I could breathe and my nose was running off that temporary congestion. I slowly went to stand steading myself against the dizziness from his blows and wincing at the burn from his intercourse. My scalded tissues had been ripped to shreds thanks to his cruel penetration. I could barely stand the pain of it.

I limped out of that dark space with much difficulty. I stood at the bottom of the stairwell looking up it. I groaned out in agony, starting my tears once more. I didn't think I could find the strength to climb them to the Elders' floor. I was sure to be thudded for being late to return to Master

Jonas's apartment. I took a deep breath and with all my will I got myself up those steps.

When I reached the sixth floor, I stood there sighing in relief that I had made it. I did my best to smooth out my clothing, then began to limp to my Master's apartment. Mad Max had been following me since the third flight in silence. He let out a yell that nearly sent me to the floor in startle.

"Oh, shit. You don't have Master Jonas's supper, fool. You are going to get the boy thudded for sure," He wailed.

I flinched then nearly went into a crying jag. "Fuck, I forgot it. Nein, I cannot go back down those steps. I am in too much pain. Gott damn it. I curse this place. I wish to hell I would just fucking die already." I began to gasp again, unable to deal with the beating I was about to get.

Christian walked up with his arms crossed watching me standing there about to fall into yet another crying jag. "Well, you can jump from the banister and join that Barnim or fall down the stairs like Drexel did. I know go into that apartment and tell the Vampire you plan to kill him. Then maybe he feels generous enough to murder the boy."

"Shut the fuck up, Christian, you motherfucker. You are not helping me," I yelled out while throwing a weak punch at him.

He easily backed away from my railings. "Well get to committing suicide, get that fucking tray or go get the thudding. I am sick to death of watching this crybaby shit."

I glared at him full of fury. "Where the fuck were you when that monster attacked u? Only Mad Maxx was here to help. Why the hell am I the one always doing all the work around here?"

Mad Max chuckled with evil. "Because you love it, schwuler boy. I think you should jump from the banister, you crybaby. Then at last I could get some peace and quiet around here. I tire of your long-winded suffering."

I covered my ears to try to quiet their bullshit voices, then took off back down the stairs headed for the kitchen. "I said shut up you sonofabitches. I don't have to listen to this shit. I am the Priceless, you know. None of you are the real thing. It is me that made this so," I shouted out loudly.

Christian and Mad Max rushed along behind me bothering me with vicious taunting. I screamed at them, never taking my hands off the sides of my head. Nothing I did would quiet their suggestions that I end the boy's life. It was driving me near mad.

I arrived at the kitchen no longer noticing the terrific pain in my flesh. I was distracted by those two idiots enough I no longer cared about my physical issues. The silvers dropped to kneeling and the blacks ran in every direction all around me. I stormed through them yelling at the top of my lungs at the evil tag team of Christian and Mad Max.

I didn't care that everyone could hear me arguing with the two of them. I ignored the frightened looks and trembling of the dozens of Haus inhabitants. I was only interested in getting my fucking job of collecting that fucking tray done.

The heavy-set female black collar gave me Master Jonas's order staring at me with an expression of concern. "Your Master called this in over an hour and a half ago, Mad Maxx. He has been ringing us asking if you come by to get it several times now."

I shot her a hateful look. "Do you see me here getting the fucking thing or are you blind. I thank you to mind your business. I didn't ask you for any information, nor wish to hear your voice. Leave me alone, damn you." I snatched the tray then turned and nearly knocked over several silver collars that came through the door without having seen me yet.

I growled at them. "Mover motherfuckers or Mad Maxx the Brutal will trash your asses. Get the fuck out of my way, all of you. I swear to Gott I am going to kill every cocksucker in this place. Do you all hear me? I am going to murder all of you in your beds," I yelled out while laughing near insane with anger.

Apparently, my wild wailing was getting around the Haus like all drama does. Somehow it reached the ears of Master Leo that was leaving the torture rooms. He had been headed to his new apartment. He took Drexel's place and was remodeling it, thank Gott. That Drexel had terrible taste.

Upon hearing that I was on a rampage he took it upon himself to collect the Elders' disturbed Priceless collar. He rushed with much speed to find me before I got the Guard called on myself with all those hateful threats I was yelling at everyone. I was rounding the hallway corner still cursing the whole fucking Haus to the terrified silver collars kneeling along the floors when he spotted me.

"Mad Maxx, you stop right there and kneel, you sonofabitch. Shut that mouth of yours this minute," roared out Master Leo sounding angrier than I ever heard him before.

I stopped and looked up when I heard my name called. I recognized this Master immediately. I thought of ignoring him for a moment, but the silver weighed down heavily on my neck. I dropped to the kneel, as ordered, still quietly damnation on the world and all the people that populated it.

Master Leo rushed to me and grabbed my collar. "Get up and follow me, Mad Maxx. That is a directive. You do it quietly. You say another word I will take you downstairs to the chains. I mean it."

I got up and followed him in high protocol carrying the tray while doing my best to quell my grumbling under my breath. Mad Max and Christian followed alongside giggling with much humor at my predicament.

"Got busted, fool," laughed out Mad Max with glee.

Christian chuckled, "Your ass is about to get flayed now, Maximillian. Perfect pleasure submissive? Looks like a nothing to me."

I shot a hateful look at him and started to yell back but Master Leo tugged my leash demanding I move faster. I dropped my eyes to the floor and endured the evil teams jeers of thrill at my being in hot water over this bullshit they caused.

I followed Master Leo in a sullen silence up the stairs until we arrived at the Elder's floor. He stopped and turned around to look at me. I dropped to a kneel, trembling with the certainty he would tell Master Jonas about my little rampage downstairs.

He shook his head. "Now that we are out of the earshot of the nosey, you tell me what the hell that was all about, Mad Maxx. Are you looking to be put to the yard? You know better than to run around screaming threats at the collars and staff like that. What has gotten into you? What the hell has happened to your face? Did one of them beat you up? You are covered in blood and bruised to shit. Christ, Mad Maxx, I almost don't recognize you."

I winced then took a deep breath. "Forgive me, Master. I think I may have fallen down the steps or something. I cannot recall. I left the dungeons as ordered by Master Jonas then everything is a blur. I could only remember he asked for his supper. I think I hit my head very hard," I lied.

He leaned down and grabbed my chin, forcing my face up where he could examine my injuries. "This looks more like someone beat you up, than a fall Mad Maxx. You sure you don't recall how it happened?" Concern was in his expression.

I shook my head. "Nein, I was going up the stairs then nothing until I was in the kitchen collecting my Master's tray."

I didn't tell Master Leo of the rape under the stairwell because I was unsure if he wasn't behind it. At that moment, everyone was a suspect as far as I was concerned. I refused to report anything until I had more time to think over the evidence.

I intended to kill both the rapist and anyone that sent him. If I dared to say a word of this crime the Haus would investigate it immediately. I didn't want to leave the punishment for the criminal, I assumed the assailant likely was a Dominant, up to the shady Voting Council once the culprit was caught.

You see in the Haus where almost every foul thing goes with their enslaved silver collars, there are still some things they consider to be crime. Their law says rape of a silver collar is when anyone other than their recognized Dominant takes special services by force. Only the Haus doesn't call this rape they call it theft.

If the culprit of the "theft of service" is a Dominant then that Master/Mistress is heavily fined or thudded, depending on the level of service stolen. In the case of a Priceless the Dominant could expect both a fine and ten lashes in the torture chamber.

If the black collar did such a horror, as you remember Olaf did, the Haus will put them to death. The black collars are forbidden to even look at the silvers or Dominants with affection as you may recall with Felix. To show open interest in a Haus sub can carry a death sentence or the punishment of exile if the collar is Priceless or the object of desire an irritable Dominant.

If it is the simple silver the black has attacked, then they will join their attacker in the orchard, no questions asked. If expensive metal is the victim, depending on their worth, they will be killed outright, beaten, and sent to circuit, or if High Born sold at auction.

If it is a Priceless that the black collar has raped, then that collar will be locked away in a cell until the Dominant feels confident the forbidden metal has been fully punished for their crime of "compliance." You see, the Dominants believe there is no such thing as a rape that was not "egged" on by the silver in some way or another. Ja, I know that is bullshit, but there it is.

They show much cruelty when a black is involved because the Dominant doesn't wish to partake the services of a submissive, they view as "tainted" by the touch of the "commoner." Unfair, but sadly truth.

If one silver dares to sleep with or rape another, without orders from their Dominant mind you, then both collars are beaten to death without argument and neither owner of the collar is fined since the two, or more, lost their investments equally. Unless that silver is the Priceless metal.

The rules stated if such a thing happened then the lesser silver, even High Born, would be put to the orchard, and the Priceless's punishment would be up to their Master or Mistress. This was why I was terrified when I discovered my level when I was involved in that mess with Ryker.

As you recall, he was indeed put to the yard, and I was viciously sexually assaulted, then leashed outside the apartment by Peter for my part in that forbidden affair. I had been manipulated into joining up with the High-Born silver Ryker when he double crossed Gretta and Peter in their attempts to take over the Haus. I had believed it was all my fault that the Guard had come for him.

I, by this time, understood that no matter how much I would like to blame myself for his death, it was not my fault. Nor was it his either. Sure, I said what I did to Felix about ending him, but I was going mad, and I had no idea the Prince was playing a role. He had been commanded by Gretta and Peter to mislead me into believing him, my friend.

His misuse of that friendship was masterminded by the brute Ben, that himself was involved in the plot. His Mistress Cora ordered him to play the bully to keep me isolated from all the other trainees or silver subs. The idea was this would force me to become lonely enough to listen to the Gretta controlled Ryker and to my then Master Peter.

The two submissive boys had figured if this plan could work for the scheming Dominants, then perhaps, they could use the Priceless powers to improve their own sorry lot. They decided to attempt to gain my loyalty to Ryker only. He would trap me in my metal, then when he broke his own – all High Borns are assured that, but no other silvers are – he would own the Priceless. Ben was assured Ryker would purchase his collar as well and reward him with a lifetime of ease and guaranteed old age at his Master's feet.

Of course, this silly and outlandish plot was easily uncovered by Cora, Gretta, and Peter. Ben was painted black for his part in this thanks to his only suggesting the idea but never getting any benefit from it. Prince Ryker on the other hand had tasted the Priceless. He was doomed for it but that was not the real cause of his ordered death.

Ryker was killed for two reasons. He had overheard Gretta discussing the secret of Peter's relationship with me, and he had betrayed the Dominants by attempting to mislead my shattered mind to gain power in the Haus for himself.

The boy Ryker was only a twelve-year-old. His aspirations for high status and control were merely those of the fantasies of the child. Peter and Gretta had no right nor business killing him for being the foolish youth. I will never forgive either of them till they die for their cruelty to a little misguided boy that didn't know any better.

Master Leo frowned at my claiming loss of memory for the violent marks and injuries to my face. "Mad Maxx, this desecration of your beautiful skin I cannot just sit back and keep ignoring. These foul idiots like Xavier, Barnim and whomever cut you up before you even got to this Haus, were the fools. Now you're falling down steps. At this rate you'll be disfigured in only a short time. If you were mine alone, I would destroy anyone that dared to leave more than a superficial bruise on your flesh. A Priceless is to be adored and enjoyed with all gentleness to preserve the perfection they possess. I will come with you to see Master Jonas this minute. I will not allow him to thud you for an accident which has so badly punished you already. I find it amazing you are injured this deeply yet still your only thoughts were to serve your Master without a single complaint. Amazing. That is the level of the forbidden silver I suppose but it still takes my breath away." He stroked my swollen cheek with much adoration in his eyes. Huh? He was gonna help me appease the furious Vampire? That was a bit unexpected, I admit.

He tugged lightly on my leash. "Okay, let's go see if we can get Master Jonas to let this late arrival slide. I am sure he can deal with his supper being cold without pulling a muscle from his thudding our beloved Mad Maxx." I got up and followed Master Leo to the Vampire's door.

Master Leo knocked lightly. Master Jonas answered immediately, appearing wide eyed and distressed. He stared at my Master Leo for a moment then noticed me standing behind him. I was keeping my eyes to the floor. The Vampire let out a cry and rushed at me. I yelped and cowered while trying to cover my head, dropping his supper tray on the floor like a dumbass.

Master Jonas ignored the spilled food. He reached down and pulled me up. Then he tightly embraced me while letting out his breath loudly as if in relief.

"Christian Axel, I was scared to fucking death. I called down to the dungeons, the kitchen, the torture chamber, no one had seen you. I found Annette gone and called her Master. There was no answer. I called the door Guards, and they said the girl left over an hour ago with papers painting her black. I thought maybe you were upset at her release and did the unthinkable." He kissed the top of my head shuddering, appearing honestly upset over the idea I had hung myself in some closet around the Haus.

I shook my head. "I apologize for worrying you, Master. Annette was painted black. Seriously? You are not just saying that after sending her away?" I pretended to be ignorant of this news to keep him from discovering my hand in her release.

Master Jonas let me out of his embrace then looked at me, his smile of thrill fading as he noticed the sorry state my flesh was in. He grabbed my chin with force, letting out a loud wail.

"What the holy hell. Who did this to you, Christian Axel. Speak up this minute." His dark eyes lit up with fires of fury.

Master Leo, who had been silent during this entire exchange spoke up "I found this collar wandering the hallways yelling with much confusion at everyone Jonas. He says he may have fallen and hit his head. I questioned him and he says he has no memory after leaving the Dungeons as you had ordered."

Master Jonas shot a look of irritation at Master Leo. "You found him wandering and yelling? What was he saying?"

Master Leo shrugged. "Mostly "shut up" and threats to murder the other collars. I must confess most of his words were nonsense, Jonas. They were just statements without a start nor clear end."

The Vampire groaned, then looked back at me pushing my sweaty hair out of my busted-up face to examine the wounds. "Did you fall, my love, as Leo suggests, or did someone beat you up? This damage looks to me more like the hand of an angered Master rather than the collision with a floor. You answer me honestly. Did you run afoul of some Dominant in the Dungeons?"

I shook my head. "Nein, Master. I swear no Dominant beat on me in the Dungeons. I must have fallen down the stairs. I cannot name anyone that could have done this. I apologize that your supper is cold and that I am late. I beg your punishment for my failure to serve your commands to perfection." Well, I wasn't lying. I didn't know the identity of the one that beat and raped me.

Master Jonas shook his head. "Punishment denied, Christian Axel. You fell. That happens. Thank Gott you're only bruised up and your neck not broken. I will call down another order to replace this one and send a black collar to get it. You come inside and lay down in our bed. You have not rested enough. That is likely why you got clumsy in the first place, ja?"

I nodded. "Thank you for the mercy and your wisdom, Master." The Vampire pointed to his hallway and bedroom.

Master Leo let go of my leash but only after I stopped for a moment waiting in front of him. He stood there staring at me, appearing not to realize I couldn't go anywhere long as he held it. Then he looked down at his hand and dropped it with a startle, mumbling an apology for his oversight to Master Jonas.

I went right to the Vampire's bedroom grateful that I had not been thudded, thank Gott. I headed for his washroom and stripped out of my clothing quickly. I didn't waste any time getting into the scorching hot shower. I wanted to scrub away that foulness of the rape, and the blood from the beating. I felt empty inside my heart. The hollowness of it brought me to extreme despair.

I was so happy to hear my Annette had made it out the Haus door. I closed my eyes imagining the look on her face when the Black collar Mistress cut that silver from around her throat. I imagined her beautiful skin lit by the rays of the late day sun and the breeze playing in her blond locks as she stepped out the Haus doors for the last time.

I bet the smile on her face would have lit up any darkness. A sudden cold rushed up my spine despite the high heat of the water running down my ravaged flesh. Without her in the Haus I was all alone, without a single friend nor anyone to genuinely care about me. I wasn't nothing to anyone, just a hole for them to use or a mouth to pleasure them. The tears began to flow despite my best attempts to stifle them. I was just so miserable.

I dropped to my knees in the shower and hid my face in my quivering lap. I let it all go without bothering to stop it anymore. I wept hard with pain in my chest threatening to cut me in two. I made no sounds. The whimpers and gasps of this grief seemed to move backward in my throat rather than coming out. My head began to ache, throbbing with each beat of my agonizing heart. The flesh became numb to the scalding water sending me to a strange place in between living and dead.

I shuddered and rocked there in that manmade rain thinking only of the rising agony within. Then with a clicking sound, Mad Maxx and I were thrown from the boy across the room. We both smacked into the wall. The force of that brain explosion threw the evil team to the floor. All of us looked at each other in shock, then back to the flesh. It rose from its position of despair and looked at all of us with anger.

Der Hund growled out, "Enough of this torment. I cannot take this pain. Maximillian, you will end this suffering now. Mad Max and Christian, you will stop taunting him or feel my wrath. Mad Maxx, you will shoulder this burden more actively with your brother. This shit continues and we will be down in that dungeon eating that horrid Helga for all our meals on a leash and in the jacket. There is no call for service so Mad Max, I call on you to attend the boy. The rest of your will sleep until he calls you for aid. Fatigue is not permitted in my shards. Move your ass, sadist, or I make you sorry for the first time in your hateful life."

I watched my brothers fall to immediate slumber. I stood up and jumped into the boy as Der Hund exited. His anger was at the level that when he leapt out, he knocked over the soap dish, shampoos, and other hygiene articles from the edge of that tub. I had seen him angry, but this was beyond anything I had ever seen before. Der Hund was demonic thanks to the crybaby Maximillian's actions. Shit.

I cleared my eyes of the flowing water and watched our Core storm from the room still railing in intense fury. I cast my eyes to the bundles of nothing that represented my brother shards lying about the room snoring.

"Told you, Maximillian. Good going fool. Instead of weeping like the little girl you should have been tracking that bastard that did this. Crying don't fix nothing." I turned off the facet and got out of the tub to dry off when the door came open.

Master Jonas stood there looking about the room, damned weirdo talks to wall. Did you know that? "Who are you speaking to, Christian Axel? I see no one in here." He glared at me with suspiciousness.

I shook my head. "I speaking to no one Master. I was singing in the shower."

He shook his head. "Well, you must listen to some strange music these days. I never heard a song so awful if that was the way it is to be sung. I thought I told you to go to bed, not take a shower. How dirty could you be?"

I began to dry off the boy. "I had neglected this duty last night then again, this morning. I thought I would clean my wounds then reapply the salves for faster healing, Master. I beg your forgiveness for not attending your command faster."

The Vampire crossed his arms. "I have been letting many slights and behaviors of insolence slide lately with you. I think you're getting used to it. I will let this go but this will be the last time I turn a blind eye to your disobedience to my direct orders. When I say go to bed, I mean go there without taking detours to showers. I am about to order food to replace the tray you dropped. Do you wish to eat something?"

I shook my head while putting back on my breeches, "Nein. I am not hungry. Thank you for asking Master. I go to bed as you ordered and stay there." I pushed past him grabbing the rest of my clothing to fold and put next to the bed before that crazy man got any more pissed at me.

I crawled under his red comforter while he watched me like the creep he is. He then came over to me and kissed my forehead bidding me a goodnight. I kept my eyes on him while he left the room and closed the door.

The second he was gone I got back up. I wasn't tired. This man couldn't make me fucking sleep. I went to his window and looked outside noticing that I could see the sunset from his vantage. I thought that it was very cliché that this Vampire would select the apartment that would let him see the sign of the coming darkness of night. I scoffed out loud while I crawled onto the window seat watching this natural display of incredible beauty unfold.

Don't tell my brothers, or I will thud the shit out of you Meine Liebe, but I pretended I was next to my Annette. I knew wherever she was at that moment she had stopped everything to watch this glorious moment. She was free at last, and far away from the hellish nightmare of the enslavement she had known all her life. No one that had suffered such a fate would miss this symbol of the end of oppression. Of that, I was sure.

I smiled as I imagined we were holding hands while we watched this special day end together. For that moment I was with her in the green fields with our Hund sitting alongside us. I could hear the baby sheep and feel the breeze on my face. She giggled thanking me for making all her dreams come true. It was the best feeling in the world when she leaned over to kiss me with her gratitude.

I was pulled back from my fantasy when my manhood hit the end of its cage causing much pain. I groaned, then leaned into the wall of the window box trying to readjust my cock. Fuck, these monsters wouldn't even allow me a Gott damned lustful thought. This was pure bullshit. I wondered if there was a hex wrench somewhere among the Vampire's things.

I got up to search his dresser when I felt a hand on my shoulder. This sent me in terror to the floor. Earlier that day, I was a little jumpy, you know. I turned around to see who my assailant was this fucking time.

To my shock their stood the shard Max. I had to rub my eyes several times to be sure I wasn't dreaming this. I gasped in pure disbelief.

"I saw Der Hund destroy you brother. How can this be you?" I stood up slowly never taking my eyes off the forgiveness/love shard.

Max smiled then said, "He did brother. Annette brought me back from the grave and you just gave me form once more."

I shook my head not understanding this nut. "Huh? I cannot build a shard, fool. How can I give you anything but a taste of my cane?"

He laughed with much humor. "Ah, you aren't so tough, brother. You're the one of us that fell in love with Annette in the first place. You coupled with her. It was your idea to save her. Not once but twice. You didn't let Christian kill her. You trusted her. It was you that got the form that released her from being found soiled. In all this, it was you that forgave Annette for joining the others in teasing us back in the laundry and washroom. Your words to her outside her apartment that day in the hallway were heartfelt and out of character for our sadist. Your poetry released her from pain rather than causing it. Then today you let her go with selflessness. You didn't even attempt to knock Maximillian out of the way to say a final farewell to the girl you loved more than any of us. Ha, just now you imagined the joy you gave her without any jealousy nor regret. When the sadist of us all can find tenderness in his heart, then I am reborn. Der Hund has his soul back thanks to the most unlikely hero, his sadistic shard."

My eyes went wide at that bullshit right there. "Ficken dich, Max. I did nothing. I saved that girl so I could fuck her. No other reason. I knew she would couple with me if I talked those pretty words, that was why I said them. I wanted to kill Stefan because he was a dirty child molester. You read me all wrong, fool. I am not the tender-hearted shard you seem to think I am."

Max laughed hard at that. "Oh, okay if that makes you feel better playing the tough guy even when I grant you great honor then so be it, Mad Max. I say to you I live because you just shattered into two. The evidence is speaking to you no matter what your lips say."

I was now a bit more than irritated "Whatever you say, pansy. What do you want from me? Oh, I know. Your will help me find that hex wrench. I take off this chastity device and you will drop to your knees to suck my cock. Get the fuck away from me. Take your nutball stories to sell to another who will buy them. I, Mad Max, don't believe a fucking word of it. Hey, wait, this is Christian's doing, ja? A joke perhaps? That cocksucker is capable of such bullshit no doubt. Well, you tell him, haha, very funny. I am not falling for it."

The forgiveness/love shard stopped laughing. "This is not a joke, Mad Max. I belong to you and wherever you go

I must follow. Now move aside. I am coming in to join you."

I flinched at that. "Bullshit. Get the hell away from me." I took off to run for the bathroom and wake up Christian to help me by getting his brother shard under control.

I didn't make it though. Max got into the boy and grabbed my leash, melting his hands to it. I was now trapped with this idiot attached to me. A nightmare has come to life, I tell you. I still have the fucker following me around to this day. I try to have a little fun, then he comes along and offers kindness to all my victims. He keeps me from doing all the evil I desire by making payments for my torture. All because I wanted to fuck Annette.

I frowned then interrupted him, "I think Master you are wrong. I say that with respect. Max was right. You loved Annette and now you love me like you did her." I winced waiting for his swat, this was Mad Max after all.

He glared at me a moment. "You too, Meine Liebe. Well, that is pure bullshit. I love to torture you and I love to fuck you. That is all. You and Max read too much into me. I am a soulless, heartless fiend, little one. I would kill you in my chains if I could."

I smiled then turned around to look at him.

He stared back hard appearing irritated. "What the hell are you looking at?"

I leaned forward and kissed his forehead. "The only person I will ever love, Master. Thank you for the Simon, Scooby and all the wonderful things you have done to help me find my peace like you did Annette."

Mad Max gasped and trembled, then for a moment started to reach out to hug me.

I saw rain behind his eyes as he suddenly remembered himself. "I did nothing. Where the hell is my cane." He was holding it.

I giggled. "In your hand, Master. I ask your punishment and thank you for the mercy of it."

He growled then swatted me hard. I yelped but turned around still smiling. I finally understood Mad Max was only the skin that holds Max's beautiful, untarnished soul.

Mad Max scoffed. "If you are done insulting me, I will continue this story, Meine Liebe. Maybe I stop telling it and beat you to the floor, perhaps?"

I nodded but said nothing.

He growled. "Ficken mich. Everyone is against Mad Max, even the fucking Kinder don't fear him the way they should. We'll see about that tomorrow when I brand you for my own. I beat you, then use the electricity too. After all that, let's see if you still love me."

I nodded but kept my mouth shut.

"Gott damn it. Fine, I tell the fucking story then. Shit, you are no fun at all, Meine Liebe. Where was I," He yelled out sounding pissed.

I shrugged.

He snorted. "I will get you. Sooner or later, you will interrupt me, then thud city for your ass. Anyway, as I was saying…"

So, there I was stuck with this cocksucker holding me back from my pleasures. Worse he claimed I created him with my "true love." What a crock of shit.

I knew when my brother Christian awoke from his slumber, I would never hear the end of his taunting me over this. I tried to struggle and get Max off my leash. The shard wouldn't budge a fucking bit. Nothing I did would release me from his hold. I paced back and forth trying to figure out how to escape this fool.

Master Jonas came to his bedroom for slumber many hours later to find me agitated like this. He frowned and had me grant him bath service, then prepare him for sleep. I was beyond irritated that I was forced not only to deal with Max staring at me with his goofy grin but doing the job of Maximillian. Damn, that is what I get for wanting to get laid by a woman for a change.

My Maser insisted I join him in his bed for his cuddle. That was gross as hell. He thought this would help me find rest, he said, yuck. I could do nothing about it but lay their feeling I may puke as he kissed the boy's neck and spoke his disgusting words trying to romance me. Double yuck.

At least he let me keep my breeches on for a change. He, however, was completely unclothed. I endured his rubbing on me till I was sure I would die from the nausea he was inciting. The second he went to snoring, I slipped out of his grip and took to pacing the floor again. I felt like the caged animal ready to rip apart the first sonofabitch I could get my claws into.

That was one long night. When at last the sun rose, I was worn the fuck out. I needed to get a break from all this stress. I went into the bathroom and called for Maximillian and Mad Maxx. They immediately awoke and stared at me with dumb looks on their faces.

"What the fuck is wrong with the two of you. Get the fuck up. You got seducing to do, don't you? I held the flesh all night. Your turn fuckers," I growled out in irritation, ready to get the hell out of this nightmare.

Maximillian, that cocksucker, spoke first. "Mad Max, is that Max holding your leash, brother? I may be hallucinating?" He looked at Mad Maxx that nodded he saw the bastard too.

I groaned. "Max says I somehow sharded outward to create him. He thinks I have such power. He is the dumbass. He believes we are a part of each other, the crazy pervert. No worries. I will take this shit up with Der Hund, so don't either of you say a fucking word or I will shatter you. I swear it."

Maximillian shook his head, appearing in deep shock. "No way, this cannot be. I have seen it all. Mad Maxx, our brother the sadist, is the soul bearer. Now I wonder about our sanity."

Mad Maxx scoffed, "Breaking news, brother, we have schizophrenia. We are shattered. Anything is possible with this shit disease."

Maximillian shot a look of hate at Mad Maxx, "That is a lie. We do not have schizophrenia, brother. Peter told us that to fuck with our head."

I nodded. "I hate to agree with Maximillian because you know I hate that fucker, but he is right Mad Maxx. We are not sick. They all lied to us."

Mad Maxx groaned. "Christ we are so fucked. Wake Christian, he will tell you I am not wrong. He knows. So does Der Hund. You two are the only idiots that haven't been able to realize this fact yet. We are insane."

Maximillian sneered at Mad Maxx. "You say whatever you like, freak. I am with the sadist on this one. This is the booby hatch we live in. We are the only sane ones in this place. Now get your ass up and join me in getting downstairs. You and I have work to do. Heidi ain't gonna seduce herself, fool."

Max, that rat bastard, and I rushed from the boy while Maximillian and Mad Maxx jumped inside. I fell immediately into a slumber, worn the hell out from one of the weirdest nights I had ever had till this day.

I looked into Master Jonas's bathroom mirror. The boy's face was bruised, cut, and swollen to shit. That rapist had made a bad appearance look worse. I wasn't sure I even owned enough makeup to make that horror staring back even look human.

I glanced at Mad Maxx. "What do you think? Would Heidi notice a face mask?" I poked at my near swollen shut left eye with a wince.

He groaned. "Ja, she would. No matter, the bitch loves the thud. She may appreciate the beaten look, ja? Otherwise, do what you can with the bases and powders. Remember yellow undercoats hide redness and peach covers the black."

I growled. "I know my fucking makeup, brother. Stop preaching to the choir. I will go stark white this day to hide all the shades of this painful rainbow but the swelling, which is too bad for us. There is no concealing that shit."

I left the bathroom then quietly knelt next to my Master's bed. "Master Jonas, I request I be released to go to the kitchen for breakfast and my appointment for training with Egon."

The Vampire stirred without opening his eyes. "Ja, ja, go my love. Be back in two hours. Don't be late this time or I will beat you for it. Let me be, I wish to sleep more." He didn't have to tell me twice.

I rushed to the room with the grey door and selected an easy to remove blouse and breeches. I left the long coat on the hanger but gave my Annette's photo in the pocket a kiss for good luck. Then with much speed and skill I did my best to cover the horror of my many head blows.

I rushed down the stairs watching the clocks on the walls with fear. I was nearly late thanks to that drawn out concealing job. I ran down the hallways like the madman sending silvers sprawling all around me in total terror. I even accidentally knocked one of them over in their kneel trying to push past him. I yelled out an apology but didn't stop to aid him back up from the floor.

I managed to catch Mistress Heidi just before she made it to the dungeon stairwell door. She looked up to see me barreling at her as I had the morning before. I fell to my knees sweating like a whore in church out of breath and wheezing from my wild racing.

She appeared flattered that I appeared so eager to chase her like that. "Mad Maxx, I had nearly forgotten you (liar). Well, you are a man of your word. Here you are."

I gasped out, "I am Mistress. I come here to bask in your glory as I promised."

The ugly Dungeon Mistress chuckled. "Ja, you have made a believer out of me. Now I will keep my promise and make a believer of you. Get up and follow me. I intend to beat the hell out of you for daring to bother me.' She reached down and grabbed my leash pulling with much force.

I got up then took my place behind her without argument trailing her in high protocol. She looked back several times, appearing surprised I was not expressing fear nor seemingly upset in any way.

When we arrived in the dungeon halls, she stopped then addressed me. "I am taking you now to my private chains, Mad Maxx. If you desired to beg mercy this is your last chance."

I shook my head keeping my eyes to the floor. "It is my pleasure to grant you your own Mistress. Thank you for your adoration though. I am unworthy of it."

Mistress Heidi snorted. "This must be a joke. No fucking way the Priceless really desires this old biddy. I could never be so lucky to catch the eye of the forbidden silver. I demand you cut the bullshit right this minute Mad Maxx. Who sent you to torment me with things I could never hope to have. Tell me the truth."

I frowned. "I think Mistress I should cover my ears. I cannot bear to hear my Goddess speaking such unkind words about herself. I am the one both lucky and tormented by your glory. If you don't want to be my lover, then send me away this second. I must beg mercy to be released from such teasing of my boyish heart by a seductress of your vast experience. I am helpless in your embrace."

That old Frau's mouth hit the floor. "This must be a dream. Wake up Heidi."

I looked up with a smile. "If you are dreaming so am I. I beg you never wake up, Mistress. Take me for your own and I will show you the powers of the Forbidden metal."

She swooned a moment then caught herself. Nein, this is a trick. I will beat the name of this cocksucker that sent you to torture me with beautiful lies.' She jerked my collar hard, then stormed to her private torturing cell with me having to practically run to keep up.

She pulled me inside the cell and slammed the door. Chains hung from the ceiling of that rock hole with a wooden table laden with thudders and tools of torture. I pulled away from the wheel and let Mad Maxx take the boy's mind.

She demanded I strip. I smiled coyly while I did this without hesitation. The harpy never took her greedy eyes off my flesh, appearing to enjoy the show. Once I was without clothing she rushed to her table and found leather cuffs. She demanded I put my arms out.

I continued to shoot flirty smiles at her while she affixed the bracelets with trembling hands. The elderly Mistress clipped me into her chains with both my arms spread wide and looked over the boy gasping loudly.

"I cannot find anywhere to thud, Mad Maxx. You are already beaten to hell. I even see healing hook wounds and other cutters. Gott damn, you are more marked up than when you come to this dungeon. I didn't know anyone could have this many scars and still claim life.' She came back around her eyes settling on my chastity device.

I grinned at her feigning embarrassment. "I apologize, Mistress, for my greed. I tend to beg a beating too often I suppose. I just love it so much. I would ask you to ignore the fury of those that did this without joy in their heart. I bet a trained hand of your legendary caliber can leave me her stripes of love despite the canvas already painted purple and black."

Mistress Heidi smiled back. "Ah, a masochistic Priceless. Now that is a dark fantasy come true for these old eyes. I cannot believe I missed this when you were here under my foot."

I nodded while openly flirting. "I was only an eight-year-old boy back then Mistress. I am near a man now. I found my pleasure, but now can you find yours?"

She went back to the table and picked up a tawse. "I am sure going to try, Mad Maxx. I will say to you with that chastity device on I wonder how you planned to bring me any thrill?"

The Mistress come around me to begin her torture. "You have no imagination Mistress if you think that cock is the only tool in my employ for taking you to your apex. That caged animal is not of much use for the woman of my dreams to truly find her orgasm anyway. Not being able to use that instrument would be my pleasure denied rather than your own. I swear to you I know the secrets of the woman well. Her lust lies inside her but not within a hole like most men seem to think. The way to thrill for her is in her mind first. Then I must capture her heart. If I can be so lucky to get that far then would be so bold as to stimulate the top of her triangle of womanhood. My female lover is then in heaven and will take me with her in her gratitude."

She chuckled at that. "You know, I did doubt that you were the lover they claim you to be, but you hang in my chains. I hear you say that you realize your cock is not the device a lady needs to find her full ecstasy. I hear you repeat the truth of all female's orgasm and the ignorance the males share. I do not recall your exact age, but I would guess it to be illegally under fifteen. Only a boy, though barely, and you have discovered all you ever need to know of the wooing of the heart of a lady. I will beat you with all I have until you confess that you were sent here to make sport of me.' She landed her first vicious blow, and I gasped in pure agony but didn't scream.

"Who sent you, Mad Maxx,' She yelled, then struck me with much force.

I yelped then began trembling. "My heart, Mistress. I love you."

She growled then hit me even harder. "You are a fucking liar. Who sent you? Tell me and I stop the torture."

I grabbed the chains that held me and sucked in my air doing all I could to beat back the bite of her thud. "I die a happy man if you beat me to death, my love. I thank you for the mercy of your attention, Mistress. I am unworthy of such adoration at the end of your chains."

For well over twenty to thirty minutes this old bitch thudded the holy hell out of the boy. She used the tawse, the cane, the quirt, and then finally the crop. No matter how many ways she questioned nor paced her blows my answer never wavered. At last, she tired of her attempts to get me to admit I had been sent.

She stopped hitting me then walked around to face me. "I have taken an inch of skin off your backside, thighs, and forearms, yet still you say you love me Mad Maxx?"

I panted and trembled but managed to put on a coy smile. "I do. If this pleases you, I come here every day if only you swear, I can bask in your adoration. I wish to call you my lover, Mistress."

She closed her eyes and grabbed her chest. Then to my shock she grabbed the back of my head and pulled me into a deep passion filled kiss. I pulled away from the wheel handing the boy to Maximillian after enduring her pawing and tongue explorations for several minutes. I was waiting to be sure this vicious Frau wasn't going to swat us any further but let me say fucking yuck.

I kissed the ugly woman back with feigned eagerness. She moaned while running her ancient hands all over the flesh. I pulled back from her kissing with difficulty when she grabbed my chastity device because she was chasing my mouth around and I couldn't escape her in the chains.

"My love, please don't torment me so. The device doesn't grow with me. I took care of my own lust earlier so I could focus on yours. Let me out of these chains and I can show you why they level me Priceless," I whimpered when she lip locked me ignoring my plea.

Then just when I was sure she would not listen to my begging her not to molest my manhood at all – ja, she wouldn't stop grabbing that damned device; holy hell the lady was persistent – she relented. I almost let out my breath in relief when she dropped her thudder and began to unclasp me. Not that I was in a hurry to keep my promise to thrill her, but she was chaffing my cock and hodensack to agony.

Once released she began a fresh attack with her mouth to mine. She pushed me with all her strength to the wall. I endured another few moments of her wanton molesting then decided enough of this bullshit. I reached down between her legs and fondled her own parts. The woman spasmed and gasped out in surprise.

I disengaged our kissing with an evil smile. "This would work better if I had you in a proper position, my love."

She nodded with her eyes wide then grabbed my leash. She dragged me across the room to her wooden thudder table. With a single swipe she knocked all the tools to the floor. I almost giggled when she sat on the table, and it groaned loudly under her weight. I thought it may break.

"Here, this will do, ja?" Mistress Heidi pulled my leash harshly near sending me face first into her lap.

I nodded. "Ja this will do, Mistress. Lay back and relax. I do all the work this time." I dropped to a knee and lifted her knee skirt. Like her twin she wore no panties. What the hell is width these old hags.

With all the skills the girls in the washroom taught me I went to work on the old hag. I spared no talents of my tongue, lips nor fingers. She panted, quaked, moaned, cried out and wailed for several minutes while I employed my oral techniques.

Then she grabbed my head by the hair and near smothered me in her ample folds of flesh in a screaming orgasm. I gasped for air praying that she would recall I required air to survive.

As her sister did the day before she announced she was "cuming" loud enough for the residents on Mars to hear. The Mistress released my mane, allowing me to pull away to catch my breath at last. That was a close one, let me tell you. I think I turned green during the act but blue from the conclusion, ja?

Mistress Heidi giggled with glee and stared at me with satisfaction glossing her eyes. "You do love me, Mad Maxx. I never thought I could be that girl that gets the handsome boy's heart."

I shook my head. "You have all the slaves at your beck and call, Mistress. You didn't need me to bring you such a release."

She nodded. "Well, that is true, but I only enjoy beating them. They fulfill my blood lust, but they are only children. Unworthy of anything but that metal I put on them or the grave they find when they insult me. These low insects around here are not good enough to fuck or fuck a Dominant of my status. I hate to admit I thought of ending you more than once in my chains. Insolent little bastard that you were. I am grateful I didn't do what I have done to others that have done less. I confess you were so damned beautiful to look at I never could bring myself to end you. I also enjoyed hearing your screams and thudding you far too much. I realize at last old Heidi's hand was stayed by the unconscious understanding you were quality most sublime. The Priceless collar of legend and you are all mine.' She leaned forward swooning.

I smiled at her full of the demons of madness. "Ja. I am and you are all mine, Mistress, at long last. I will never forget your torture. It has kept me awake many years. I swear I cannot count the number of nights I close my eyes and see your face in the darkness. I have returned to you for an equal return of your adoration for me, and all the many children you have culled to their graves and broken their collars. You'll find in my eager embrace a place you truly belong."

The Dungeon Mistress reached out and played with my sweat drenched locks. "When will you come back and take me in your arms again, my love?"

I stood up wincing from the foul injuries I had suffered in my effort to seduce this killer of children. "I will return to you in the morning. Can you meet me at the top of the dungeon room steps, Mistress? If you say ja, then we can come here and do this all over again. This time wear something easy to remove and sexy. I would like to show you other powers I possess." I winked at her with a wanton smile.

She nodded wildly. "Ja. I can be there. At nine thirty I will be on the top step waiting to explore the powers of the forbidden metal."

I redressed quickly while the Mistress watched with silent lust filled eyes. When I had completed my task, I walked back to her and kissed her cheek.

"I must regrettably run, my love. I see you in the morning, Mistress. Promise you will dream of me tonight? I wish to know you cannot get my image out of your mind." I grinned ear to ear as I retreated walking backwards.

She nodded. "I will think of nothing else I can assure you."

I turned then fled from her torture cell headed for the desk of her sister the Dishonorable Helga. I found the ugly Frau there as always. She saw me and dropped her magazine with a huge smile.

"Mad Maxx, you came back.' She looked around to make sure we were alone.

I smiled. "You're all I can think of my love. Of course, I came back. Tell me did you dream of me?"

The ugly Mistress Helga nodded. "You damned right I did. My classes are unruly as hell from all my daydreaming. I swear I near killed the wrong silver this morning in my confusion. I need you to take care of my desires or I will have to put down my thudders forever.' She giggled over that.

I chuckled full as the demons within rose again. "Ah, well you are in luck meine beauty. I have the morning off tomorrow. If you could meet me at the dungeon steps at nine thirty-five wearing something sexy and easy to undo, I can finish what we started, ja?"

She gasped. "Oh, meine Gott, Mad Maxx, my sister Heidi will be coming back from breakfast at nine thirty. She will catch us."

I frowned, "Vell shit. Can you not hide out and wait till she clears the steps? I have only a few hours, barely enough time to even begin." Mistress Helga yelled out interrupting me.

"A few hours are not enough time.' She grabbed her chest, her eyes wide in shock.

I chuckled with mischief. "See that is why the Priceless legends exist my love. I know how to take your breath away. When I am done you will never desire another pleasure. I will satisfy you until you're sure you have died and received the eternal reward you deserve."

Mistress Helga nodded with her mouth open wide. "That sounds heavenly Mad Maxx. I will be there. I wouldn't miss this demonstration of the forbidden metal for my life."

I walked behind her desk and pulled her into a deep kiss, enjoying that she didn't realize her sister was on my breath, then pulled back. "That is all I ask. I see you in the morning, my love." I took off down the hallway leaving the woman panting in her desk seat.

I shot a smile at Mad Maxx. "Great work, brother, but let me say I don't appreciate that we will not be sitting for a fucking week over this double seduction."

He chuckled. "Maximillian, I am beginning to agree with my brother Mad Max that you are a crybaby. Damn, we cannot have me siding with that sadistic bastard so cut your bitching out."

I nodded. "Ja, okay, but next time you get your freak on, leave me out of the boy. Yikes, you're a fucking animal brother."

I was busy chattering with Mad Maxx and rushing down the hallway looking at the floor I didn't see Peter. I smacked right into the bastard without even a bit of slowing down.

He growled out in fury, "Gott damn it, Maximillian. Look where you are going, you little bastard."

I jumped back, then fell to a knee blinded in my memory for a moment by fear. "I apologize, Master. I beg your mercy." I recalled too late the Peter no longer held my collar.

He stood above me chuckling at my reflex behavior that he had beaten into me. "Now that is better. What the fuck are you doing down here? Answer me submissive."

I shook my head. "Nothing that would matter to you, Master Peter. I come to look for Master Leo for my Master Jonas."

Peter nodded with a scoff. "Well, I come to look at the new crop of slaves just in before those Dungeon Mistresses kill half of them off. I am in the market for a worthy submissive since the last one I had wasn't the quality metal I thought him to be."

I said nothing to his insulting me. I kept my eyes to the ground cursing this cocksucker for all the horror he and my mother did to me. Then to stand there calling me less than quality. Fuck him.

When I kept my silence, he snorted with much arrogance. "I will get another chance at stomping down your will, Maximillian. I already know you will break that collar. I may even get lucky and punish you for your betrayal sooner. If I were you, I would be watching my back little man."

I looked up with hate in my gaze. "Why would that be, Master Peter?"

He chuckled. "That got a rise out of you, didn't it. Well, you see Maximillian, I am not amused by your crazy act in the cells. I admit you fooled me, but I will never forgive you for it. I realize now you and that Vampire plotted the whole thing. You tricked me into tossing my collar. You and that fucking Jonas. Do you really think that nothing can keep me from taking what is mine? Jonas cannot protect you round the clock. I will come for you anytime I am ready. I will make you pay, slow and painfully, for playing me the fool."

I frowned at that. "Wait, you don't think me schizophrenic, Master? I don't understand. You had the Guard called and tossed your collar saying I was mad."

Peter rolled his eyes. "Cut the act, Maximillian. I am not going to fall for this shit twice. Ja, I called the fucking Guard and tossed the collar when I thought you daft. I know what you did, don't try to deny this. You are as Priceless as the fucking backstabbing Leo is. I am tired of this conversation. Get out of my sight this minute. You got yours coming. I will make you wish the Guard buried you when I am done making you pay." He stormed off down the hallway leaving me there fully confused at his oddly accusing me of being involved in a plot with Jonas to steal my collar from him.

I got off my knees and thought about that for a moment. This had to be a mind game. I recalled Ryker warned me about these games the Dominants play with the inexperienced silver collars. He was fucking with my head was all. He hired that fake doctor, and paid Hemmel to beat me. I was there. I saw and heard the whole thing. This was merely his attempt to confuse me, or was it? Did Jonas find a way to maneuver my silver into his claws as Peter claimed? Did I accidentally aid him to do it?

A memory flooded through my mind. *"You are mine Christian Axel, even though you nor Peter know it yet"* and *"I finally have the Priceless collar I have been seeking all my life within my claws. I will never let it get away."* I heard the Vampire Jonas's voice ringing through my ears.

I staggered from the dungeons deep in thought of this possibility. Don't get me wrong, I wasn't upset to be set free from the lust of my father's bed. I merely worried perhaps Master Jonas's manipulations of me could have gone further than just that eight months of strangeness. Had he been drugging me all that time to make me appear mad?

I then thought of the death of Barnim and Drexel along with the horror of being under the dominion of four Elder Masters. Could they all be in on the plot? What the hell did they expect to gain by holding my forbidden metal other than the obvious special services?

I was deep in my focus headed for the kitchen when I passed the storage door of where Ben and I had our fatal last meeting. I hadn't passed it by a couple paces when someone tapped me on my shoulder. I turned around to see who dared to be so rude.

My face caught fire as the pepper aerosol shot into my eyes and mouth with vicious force. I fell to my knees gasping for air in a spasm clawing at my burning eyes. I kicked with all I had as the assailant dragged me across the floor helplessly struggling for air and blinded. I couldn't even tell where he was taking me.

I heard a door close and smelled the faint odor of garden chemicals just as my nose congested fully. I realized the rapist had pulled me into the closet I hated the most in all the Haus. That sonofabitch had been following. All that time waiting for his moment to strike where no one would see his abducting me.

I coughed and choked unable to do more than whimper as the monster rolled me to my face then pinned me with his weight as he did the day before. I heard him spitting as he pulled down my breeches.

I braced myself best as I could for the coming terror. I realized with great horror I had become the target of a serial rapist.

Chapter 42: Warm Heart, Hot Spray, and Cold Shower

I trembled waiting for this rat bastard to start his brutal sexual assault. He had hesitated after getting my breeches down. During this time of his holding off his rape, I clawed at the floor helplessly. I gasped and coughed unable to breathe as during his attack the they before. I heard him letting out his breath loudly but still he didn't start his intercourse.

I suddenly realized the sight of my severely bruised and cut backside was what caused him to pause his assault. It seemed he was holding me down trying to decide what to do. I took this opportunity to try to escape him. I grasped around me in my blindness looking for anything to offer me leverage or use as a weapon. My right hand came across what seemed to be the leg of a table. I grabbed it and pulled it with all my might.

It didn't aid in pulling me away from him but whatever was on the top of it spilled to the ground loudly. The racket appeared to have awoken him from his deep thinking, or whatever the hell he was doing. He backhanded the back of my head sending it smacking into the floor. I felt dizzy and sick immediately from the blow.

I felt him rolling me to my back. I coughed and spasmed blinking wildly trying to clear my vision to get a look at this sonofabitch. I couldn't keep my eyes open no matter how much I willed them. When I did get the tiny moments of lifting my lids, I couldn't see anything but bright blinding lights with horrid burning. Tears were pouring out of them by the buckets.

I felt him tearing off my pants, then he forced my legs apart and crawled between them. I kicked at him the entire time, but he subdued my struggle with some difficulty. He was strong and apparently quite fit.

His put his hand around my throat. I grabbed his arm clawing with all I had. He didn't appear affected as he used his free hand to force his entry from this position. I began to gasp and cough wildly from the terrible pain of this further assault on my chemically burned sphincter.

I punched him in his chest and pushed trying to get him the hell off me. He coupled harshly then stopped and captured my flying fists. With incredible strength he pinned my arms to my chest. after he had subdued me this way, he began his rapid, brutal thrust. I was helpless to do anything thanks to the effects of his pepper spray and his huge size. I had no choice but to lay there and take his penetration.

I gasped in agony doing my best to listen to anything he said or did that may point to his identity. I could hear him panting loudly but he was careful not to use his voice in any way. This motherfucker was making damned sure I had no clues to work with. He took his time, stopping several times when apparently about to reach his climax. Much to my misery it was clear he wanted to make it last as long as possible.

This was a clue I paid attention to. He was enjoying this act for the pleasure of the sex with me. I had at first thought maybe this was someone trying to teach me a lesson. I decided based on his behavior that it was likely the bastard had been lusting after me but had been unable to obtain a leash.

Since I was the Priceless that information didn't help me much. I had become the most desired silver collar in the whole Haus by this time. There was not a Dominant in the place that had not made an offer for a single leash with me. The only exception was Peter and my mother Agnette. Thank Gott on that one. I was already depraved enough without adding that sin, ja?

Master Stefan, Mistress Heidi, and Mistress Helga had already made this evident. All three had been in shock when I showed up offering such an unbelievable deal. Their belief, not mine by the way. My seductions were super easy because of my reputation as the incredible lover that could kill you with a single couple. It didn't help that most in the Haus also believed in the many legends of the forbidden metal. You know, adding years, youth, and vitality).

Thanks to those facts, the gender, size, and brashness of this rapist were the only things that narrowed down the suspects. For example, I could rule out any black or silver collars. Neither a person of such a low level would have taken such a risk, not once but twice.

For starters, only a Dominant would be merely heavily fined for such a theft. My Priceless level would assure death for anyone else caught doing this shit. A black or silver collar also could not have the leisure to stalk me and attack at various times while being capable of taking his time with the intercourse. They would be in a hurry, worried, and not likely to be carrying around a can of pepper spray.

I continued to endure his painful and drawn-out sexual assault by focusing on figuring out who this could be. My mind whirled trying to match up any male Dominant that could fit the physical description of that shadow. Then suddenly I got an idea.

The rapist thrust more harshly, appearing to become enthralled. He let go one of my arms to grab my waist to keep me from sliding away from his pounding. I took that moment to reach up blindly seeking his face. My hand collided with his jaw, and I ran it down his cheek then across his lips. He growled out and stopped his mount. He slapped my arm away from his head then backhanded me to near stupid.

He took hold of my hips, letting go of both my arms. With great violence he restarted his assault, holding back, nothing. It was at this point I, thankfully, lost consciousness from the agony of his brutal thrusting and his beating me for trying to see his face with my hands.

I awoke to the cold rag of a female black collar. She had come to the closet to get chemicals for her moping and found me passed out. She had a clean bucket of water on her. after checking to see I was still breathing the woman had immediately attempted to rouse me from my state of faint. I lucky came back alert quickly before she had a chance to report finding me in this state.

My eyes were red and swollen from the pepper spray, but I could see once more. I sat up and noticed the rapist had redressed me. The pain in my backside, burning skin from his spray and new bruises were the only indication he had ever been there at all. I looked at my terribly upset attendant kneeling next to me.

"What time is it? Uhm, I don't know your name." I was terrified I had passed out and was now late to get to the apartment without inciting Master Jonas's anger.

The Frau appeared startled. "Uhm, it is eleven. Mad Maxx, I am Evelynn, don't you know me? I have given you the Master's food trays every day for years. How did you end up unconscious in this closet?"

I shook my head. "Ah, ja. I remember you, Evelynn. Uhm, I am not sure. I guess I was lost or something. Wait, why are you here again? Is it breakfast time?" I was having a little trouble with the confusion. Thato happens sometimes when I am stressed.

Evelynn's expression appeared worried. "Mad Maxx, I think I will call Olaf or Vilber to walk you back upstairs. You look horrible. Did you fall maybe? I never seen eyes so red."

I heard those names and grew angry at once. "You will stay the fuck out of my business, Evelynn. Do you know what happens to people around here that mess with me?

They fall down and don't get back up. Leave me alone. Why are you bothering me? I didn't ask you for any breakfast tray." I stood up swooning for a moment nearly falling back to my ass.

The black collar Evelynn stood up too. "Mad Maxx, you are ill. I am not trying to bother you. I am trying to help."

I glared at her while leaning into the wall waiting for the dizziness to pass. "I didn't ask for your help, Evelynn. I mean it leave me alone."

She appeared frightened and looked at the door as if ready to run for it. "I will leave you to your peace then. I am leaving now, if this is, okay?"

I sneered at her, "I said leave twice, didn't I? Are you deaf? Get the fuck out!"

Evelynn rushed for the door leaving her bucket and chemicals behind. I winced, then groaned at the extreme pain tormenting from what seemed like everywhere. I was quite sure I couldn't survive too many more severe thuddings from Mistress Heidi, followed by a beating and rape of chemically burned parts. I was sure I would not recover the next attack from this pervert.

I waited a few more moments till I was sure I could walk without fainting, then fled the closet. I had another twenty minutes to return to Master Jonas without fear of his tawse. I breathed a sigh of relief that at least I had that mercy. I limped slowly down that crowded hallway turning around to watch behind me every so often. I tripped several times over kneeling silvers.

I would curse them for near sending me to a spill but that didn't stop my paranoid watching my back. I assumed that the rapist had his fill, but I couldn't be sure. Then I saw a kneeling silver I recognized. I called out to the boy.

"Abelard. You bastard. Where have you been," I growled at to trembling silver.

He looked up with a start. "I apologize, Mad Maxx. I thought you were put in the yard. I only just heard of your return from the long training you attended. I am here to serve you in any way you need."

I smiled at that. "Well, if you are not busy at this moment. I could use an escort back to the sixth floor."

Abelard nodded. "I am not busy, Mad Maxx. I am happy to come with you this minute." He stood up and began following me.

I felt much better having this extra set of eyes to watch out for my assailant. "Abelard don't walk behind me brother. Come keep me company. Keep me entertained with the Haus gossip?" I motioned him to join my stride.

I saw the silvers around us gasping and falling to their faces. I frowned at this weird behavior. I shot a look at Abelard that had caught up with me. He shrugged, appearing as confused as me at their apparent fear of the two of us.

He then smiled. "I want you to know I used to be afraid to be associated with you, Mad Maxx."

I chuckled at that. "Oh? Why is that my friend?"

Abelard laughed. "Shit, you are Mad Maxx the Brutal. I was sure you'd beat me to death or desire I couple with you, then get me shot by the Guard."

I roared in laughter till I nearly choked. "I have to be honest with you, brother. I don't like the sex with the man. You were safe. Okay, I did think of caning you a couple times. I heard you could take such a thudding no problem. That was not true?"

The silver collar nodded. "Ah ja, I can take the thud, but who the hell wants to? Hey, did you hear Annette got painted black? It is all over the Haus."

I stopped in my tracks and Abelard stopped too. "You heard this news already? How? That only just happened."

Abelard smiled. "Are you kidding? That Annette was the fox, brother. All the boys waited in the kitchen just to watch her pick up her Master's tray. I asked her to give me a blow job in the closet and she said she would do it for nothing. Can you believe that fine piece of ass is nothing more than a whore? You know, no one has seen Master Stefan in days. There is a rumor that he sits on his couch rubbing one out to the photo of his mother."

I gasped. "What the hell. How can that rumor be true? You looked at my Annette with lust in your heart. How dare you, motherfucker." I turned around and swung at him.

He let out a yelp as I slammed into the wall. I grabbed his throat to kill the fucker right then and there. I don't usually do the killing you know but this sonofabitch said that about our sweet Annette. Even Mad Maxx was for ending this nasty rat bastard.

Abelard struggled for his breath while I tightened my hold on him. "She is not a whore. Take that back, motherfucker. Abelard, start apologizing or I swear to Gott when I finish murdering you, I will bury you right where you fall."

He gasped and sputtered, clawing at my hands with terror in his eyes. "You're crazy Mad Maxx. You know she was a whore because you are one too. Instead of calling you the Priceless they should call you the pass this, taste this metal.

That was fucking uncalled for. I was beyond pissed by this time. I squeezed his neck hard with rage filling my chest like the furnace of hell.

"Annette has the baby sheep, you creep. I am going to put you to eternal sleep. They say what you sow you shall reap. Then the word whore you never repeat," I shouted out as Abelard turned blue and his eyes bugged from their sockets.

All around us the silvers fled in terror. The black collars let out yells of caution demanding someone call for assistance. I didn't care. Let them call the fucking Guard. This cocksucker was going to die. No one calls the girl we love a whore, then gets to breath the air like the living does.

Hands reached from behind me and pulled me off the smothering Abelard. I wailed out in both rage and fear. I thought maybe that rapist guy again you know, honest mistake. I turned in the grip blindly flying at the eyes of my attacker.

"Don't touch, punch much. I know who you are, star, far. I am not a whore, store, poor." I tore at the face of the man for daring to try to grab me right there in broad daylight.

The man grabbed my leash and jerked it hard sending me face first to the floor. I tried to get back up, but he kicked me in my stomach hard. That sent me rolling and gasping, unable to get up or defend myself.

Above me I heard a voice that sounded like Master Leo's. "Mad Maxx, stop this at once. That is a directive. Do you hear me? Answer me."

I groaned out. "Ja, I hear you Master," I whimpered while trying to get my breath.

I opened my eyes and saw Master Leo reaching down to regain my leash. He knelt next to me, his eyes wide and sweat pouring down his forehead. He looked frightened for some reason.

"You know who I am, Mad Maxx? What the hell? Your beaten to shit. Did Jonas do this to you for being late yesterday? Why are you attacking this wall? Who the hell is Abelard?" He pinned my flopping head and looked into my eyes.

I felt sick to my stomach. I was so damned confused. His questions made no sense to me. I wanted to answer him like the perfect submissive I am, but I didn't know how to fulfil his requests. I shook my head and groaned from the agony of all my injuries, and Abelard's foul betrayal.

Master Leo winced as he pulled my face from side-to-side, witnessing all the swelling and fresh blood of my latest beating. "Jonas has to stop this shit. I won't stand for it. Get up Mad Maxx. You are coming home with me. If that bastard wants to call me out, let him." He grabbed me by my upper arm, pulling me to my feet.

I shivered and began to feel odd. I shook my head unable to stop the rising urge to move about. I wrung my hands and rocked on my feet. Everything was moving the wrong direction. I shot a look of fear at my Master, unsure of what was happening to me.

"I have to get back to Master Jonas or he will thud me, Master. Thank you for this mercy. Is it breakfast time? Did Evelynn come by here? I think she was supposed to, wait, did you see Abelard? That bastard was supposed to be doing Master Peter's laundry. Why the hell are you repeating me? I asked you a question. Where is the laundry? Did Ryker tell everyone those lies about Annette? That sounds like something he would do. Maybe Ben, did it? Fucking Ben. Get Abelard. He is the one that deserves that thudding not me." I said to Abelard or was it Master Leo?

Master Leo was not happy that I thought he was the lowly Abelard. He reached out and slapped me hard. I gasped and grabbed my burning face cowering under my hands. He pulled my leash hard.

"Enough of this. Silence now and follow me. I mean it or I take you to the chains, Mad Maxx." I felt my leash being pulled I grabbed it in terror holding on but followed the man storming off at the other end of it.

I was lost. I didn't recognize anything. Everywhere were strange spinning lights and shimmering lines. I whimpered but held on to the leash that seemed to be pulling me into the nothing. I could hear something electrical above my head. I looked up and saw the spinning vortex of a storm of cataclysmic strength. I recalled seeing this nightmare world before. I wailed out in terror. The madness had caught me in her webs of the tapestry.

The man pulling me along stopped and slapped me again. I couldn't see his face thanks to the shimmering lights of this place of monsters. I fell to my knees and wailed louder when the floor began to spin away. The walls were melting into the static. I was losing my mind.

That man picked me up and threw me over his shoulder like a duffle bag. I couldn't figure out my location. He said something but the whirling vortex sucked his words away. I watched them fly to the sky just as he flung me over him.

I wailed and thrashed thinking he was taking me down to the hellhole below. I feared the jacket and cell, but I was helpless. I couldn't even recall my name.

Instead of going down, I felt myself floating upward. I thought the vortex had captured me like it took those words. I stifled my screams of terror sure that all was lost. I was so scared I could barely breathe. I wondered if it would hurt much to be ripped apart by that cyclone. I closed my eyes falling into a weeping jag at the hopelessness of it all.

I fell to the floor still weeping. I rolled up into a ball unwilling to fight anymore. I felt hands grab me roughly. They lifted me then dropped me again. I gasped out in shock. It was so cold all the sudden. Then I couldn't breathe. I started to struggle. The hands were holding me under water.

I fought with all my strength to get back for a breath of air. I was released just when I was sure I would drown. I took a loud gasp the second I was above the surface of that freezing liquid. Then to my horror the hands forced my head back under the frigid element. I clawed the hands holding me down, kicking and spasming out trying to get back for air once more.

Like before, just as I was losing my battle for life the hands released me. I came back up and saw Master Leo there. I reached out trying to beg him to help me. He had an expression of fear on his face. I thought he must be able to see my attacker.

I sputtered out, "Master, help. They are killing me." I was grabbed then dunked under the drink once again.

I was held there for several moments then allowed to return to the surface for a breath. I realized at this point I was in a bathtub filled with ice and water. Someone had put me in their clothes and all. They were doing the breath play, letting me get air only seconds before I would take the frigid death into my lungs.

When I was permitted to breath this time I wildly grabbed for the side of the tub. I held on for dear life sputtering and gurgling. I was soaked and trembling with the chill. I saw Master Leo kneeling next to the tub, his sleeves rolled up. His clothes were wet from the splashing of my struggles to escape his pushing me under repeatedly.

I set my eyes on him wailing out, "Please mercy, Master. I beg of you."

Master Leo frowned then grabbed my hair. "Say my name, Maxx. Say it."

My lips were stiff from the cold. "Leo. Leo, please mercy," I shouted out.

He let out a breath of relief rubbing his eyes. "Do you know where you are, Maxx?"

I nodded. "This is Master Drexel's bathroom. I assume I am in your apartment Master. Please, I am freezing. Mercy, I beg of you." My teeth chattered loudly as I tightened my grip on the edge of the tub, sure he was going to dunk me again.

Master Leo shook his head. "Not yet, Maxx. I need to know you are back with me. Now answer me. What is today? Do you know the year? What time is it?"

I shivered uncontrollably. "If not, much time has passed it is eleven in the morning. It was Thursday, and it is 1972, August I think."

He nodded. "The time is close enough. All the rest is correct. Tell me what happened. I find you yelling and attacking the wall calling it Abelard. You're beat up again. Did Jonas do this to you, or did you do this to yourself in your agitation?"

I shook my head in confusion. "I wasn't attacking a wall, Master. I beg your forgiveness, but I was angered by the silver collar Abelard for cruelly insulting me. I know it was wrong, but I lost my temper and was hitting him for it." I looked down sure that I would be thudded for my stupid move over the words of a nothing.

Master Leo gasped. "Maxx are you speaking of Gustov's silver collar Abelard? Is this the boy that you say insulted you?" He leaned forward and I white knuckled my grip sure he was getting ready to push me under the water.

I nodded. "Ja, Master Gustov is his Dominant."

My Master groaned. "Your illness is deep, my love. Abelard has been dead over six months Maxx. The boy got shot by the Guard for trying to escape. I saw this with my own eyes."

I gasped in horror. "Nein, I was just speaking to him, Master. He cannot be dead. Why the hell would he try to escape? He was a perfect submissive with great marks. This is a head game you play with me."

"What the fuck, Leo," Master Jonas bellowed out from the bathroom door.

Master Leo turned to his voice. "Jonas, how the hell did you get into my house? It's rude not to knock first."

"I did fucking knock. For many minutes now. You said Christian Axel was in a psychotic state." He rushed into the room, but Master Leo stood up quickly and blocked him from reaching me in the bathtub.

Master Leo growled. "He was Jonas. I found the boy beating up a wall. He claims he was in a fight with a collar that was put to the yard six months ago. That silver of Gustov's, the one that ran. He was hallucinating, rhyming then began shrieking. I had to restrain him then carry him up the fucking stairs. He laid on the floor out of his head while I filled this tub with ice and water. I have been dunking him for fifteen minutes to get him back to his senses. Look at him. He is beat to hell. Did you do that to him? If not, then I would like to know who did."

Master Jonas shook his head. "He was fine when he left our bed this morning. There were no signs of psychosis nor this injury I saw on his face. This must have all happened while he was out of my sight."

Master Leo scoffed. "No signs of psychosis? Maybe you didn't bother to check. Yesterday he was raising hell to everyone. Today he is attacking no one. That, brother, is fucking signs of schizophrenia. This boy is a real Priceless. He needs treatment or this will get worse."

Master Jonas looked at the floor. "He is already getting treatment, brother. Over a year's worth already. I thought Claus told you. He was diagnosed some time ago."

Master Leo looked shocked. "You said over a year of treatment? Wait, Peter knew of this boy's true nature? He realized Maxx is Priceless?"

The Vampire groaned. "He was told the boy would not recover when the forbidden metal set in him. He was hidden in the dungeons while the madness took him over in onset. Peter told me the boy had a few minor issues prior to this full cycle of the disease. He ordered the Guard and uncollared the boy when it was determined he is a true Priceless."

Master Leo shook his head. "This cannot be right. Peter told me the boy was a fake and level without evidence. Why the hell would he toss him the second he found that the collar was true legend?"

Master Jonas took a deep breath then shot me a look. "I didn't want to say this in front of Christian Axel. I paid Peter's doctor to tell him that the boy was a severe Hebephrenic that would never recover. Peter thought the collar would be hopelessly mad and require lifetime hospitalization."

I growled out. "I knew that doctor was a fake. Why did you do that, Master? I am not the schizophrenic." I said through chattering teeth and trembling lips.

Master Leo glared at me angrily. "You will be silent, Maxx. You are a schizophrenic. Jonas only got this doctor to lie about the seriousness of your illness. You are Priceless, this I see without any doubt. You stay out of discussions of your betters, or I will dunk you for the fucking fun of it. You will mind me, Gott damn it."

I looked at the floor trembling. "As you wish Master. Thank you for your mercy."

He turned his attention back to the Vampire. "This boy needs more help than this fucking Haus doctor can give him Jonas. The disease he has is fucking devastating. He is in the Acute cycle and the stress of this life he endures is setting it off to nasty symptoms. I demand you speak to Claus, and we get him sent to the professional institution."

Master Jonas looked at him with his eyes wide. "Are you out of your fucking mind, Leo? Even if such a dumb thing were permitted, he would be automatically trapped in his metal. He has to break my collar. Christian Axel will be fine. The private doctor I use says he has recovered nicely and may never have another issue with the psychotic shit but that one time. What the hell do you know about anything anyway?"

Master Leo roared out angrily, "My beloved sister was the schizophrenic, Jonas. I know a great fucking deal about this nightmare disease. I watched my best friend in the world be consumed by it for years. Then before the girls twenty-fourth birthday, she could deal with her demons no more. She ran a pipe from the family car and ended her pain. I refused to marry or have kinder thanks to that monster that runs in my own veins. Our family has seen several of its members taken to hell by it. That is how I knew to calm Maxx down with the frozen water and submersion. I almost couldn't get him back this time. You would be willing to injure this boy, even lose him to suicide, over a fucking collar? That is beyond imagination. I thought you loved this submissive Jonas. You are a fool. This boy is in real danger of being killed by his uncontrollable symptoms, either by his own hand or the Guard."

The Vampire glared at Master Leo with fury. "You are being dramatic, Leo. Christian Axel is recovered. The doctor says he is in the calm part of his cycle. He is logical and not psychotic this minute. The Acute you speak of I have seen with my own eyes not that long ago. This is not that part of his disease, trust me. You say he is sane over a little dunking in the freezing water, but when he was sick before nothing worked. I think you are making insolence into symptoms, is what you are doing."

Master Leo crossed his arms shaking his head wildly. "Nein, I do no such a thing. Even the acute schizophrenic can have moments of clarity that last for a few hours, even days. Only the onset of the disease is so severe. Maxx is no longer in that early stage of the illness. He will cycle and show symptoms that will get worse when he is heavily stressed all his life. You better face that fact this minute or be ready to pay for such a dangerous denial. He needs rest and a quiet place to settle down. Otherwise, be ready to watch your back. The schizophrenic can even kill if agitated enough Jonas. They don't even know that what they do is wrong. The increase in his violence is written all over his flesh. Loot at him for Gott sakes, brother. I have known this collar for a few years and never have I seen this much injury in him, not even when under the leash of Xavier. I warn you right this minute. You lock this boy up or you will be sorry."

I shot a look of fear at my HusDom Jonas. "Please Master. Don't lock me up. I swear it was an accident. I got dizzy from yesterday's fall and thought I could just ignore it. I compounded my mistake from that other fall and fell again. I hit my head hard this time. I remember this. Call Evelynn down in the kitchen. She can verify this. I spilled my breakfast and was retrieving a cleaner on the shelf in the storage closet. The container was up high. I slipped like a fool climbing up to reach it. I am not the Psychotic I swear it."

Master Leo turned and shot me a look of intense fury. "Did I tell you to be silent, Maxx? I think I did. You are asking for my quirt. Shut your fucking mouth. No one gave you permission to speak." He reached out and slapped my face with a nasty sting. My flesh was wet, yikes. Never get smacked when you are soaked. Hurts like hell.

I grabbed my burning cheek and looked at the floor immediately. I panted in terror that my Masters were going to haul me to the dungeons below. I prayed that Master

Jonas's desire to keep me with my program to break my metal would win over Master Leo's misguided beliefs that I was a threat to myself and the residents of the Haus, because I wasn't.

Master Jonas growled out at Master Leo. "Enough. Do not strike my man, Leo. He is only reporting new information he has remembered just this moment. I will call this Evelynn to check on his report of falling in the storage room. You come with me, and we ask her if Christian Axel is speaking the truth. If he did fall again, then it would explain all the behaviors you saw. A hard spill could confuse the boy. A mild concussion can cause temporary madness. If this black collar confirms this happened, then you will drop this discussion of hospitalization and I take him home."

Master Leo scoffed. "You can call this black collar fine by me. Then you call Claus and Bladrick. I demand this case be discussed right here and right now. Maxx goes nowhere until I have my chance to speak my concerns about this collar to all the brothers."

The Vampire nodded. "Fine by me Leo. Christian Axel, get out of that tub and those soaked clothes. You dry off and come to the living area kneel and keep silent. This is a directive." He stormed from the room leaving Master Leo standing there with his arms crossed.

I shook from the cold but didn't move from my spot, "Master I don't have any other clothes."

Master Leo glared at me. "Then I guess you come kneel naked, don't you Maxx. You are extremely sick and this man of yours will not listen. I am not trying to be cruel to you, love. I know you don't believe me, but I am only seeking the best for you. I couldn't bear to see you put to the yard or commit suicide like my sister did."

I kept my eyes to the floor. "If I go to the hospital, I am trapped in my metal Master. I desire nothing else than to be free of this subjugation. I beg your mercy to drop this plea with Master Claus to see me forced on my knees for all my days."

My Master sighed then looked at me with pity in his expression. "You don't understand Maxx. Your disease already has assured there will never be freedom for you, love. This madness will make you vulnerable and there is no cure for it. You will require someone to look out for you when the acute phases come on. Without a guardian, you're doomed to end up in the early grave. Even if you broke that collar, Maxx, schizophrenia is still your Master. That is why it is Haus law that no Priceless is allowed to use the sacred bolt cutters. They cannot survive on their own thanks to their illness. I refuse to argue with the insane. Do what Jonas told you to do. Hurry up or I will come back and punish you for your insolence. This shit will be settled shortly. Claus is a reasonable man, unlike your man Jonas. I am sure he will understand that sending you for help is in your best interest." He shook his head, turned, and followed Master Jonas from the room.

I rapidly crawled out of that freezing water. I pulled the plug just to be sure Master Leo could not easily put me back in there. I removed my sopping clothes and wrung them out best I could. I put all them into the tub to drain, including my soaked boots.

I could barely grab one of his towels from the bathroom rack, I was trembling so bad. I wrapped myself in the soft warmth of the cotton cloth for several moments grateful for the comfort. I thought of the vicious thud marks from Mistress Heidi. If I walked into the living room naked, the Masters would see the stripes. That would bring many questions.

I thought of my situation after Peter's collaring. I had covered myself with his towel to retrieve his tray from downstairs. I decided this trick would have to work once more. I wrapped myself in Master Leo's large bath towel around my waist, covering my backside and thighs. My forearms were not as badly bruised but I decided best to keep them behind my back anyway just to be sure.

I fled his bathroom rushing to the living area. I could hear Master Claus and Master Jonas already in there. The men were arguing with Master Leo. I took a deep breath and braced my nerves. I silently entered the space and rapidly fell to a kneel at Master Jonas's feet putting my arms behind me with my head bowed.

The Masters fell silent on my arrival for a moment. I felt their eyes on me. Then Master Claus cleared his throat.

"Mad Maxx is subdued, Leo. He doesn't look aggressive to me. Quiet as a lamb in fact. You dragged me out of my nap for this?" The old crossdresser snorted at that.

Master Leo growled out, sounding irritated. "You didn't see him less than an hour ago Claus. The boy was beating the shit out of the wall and screaming like a beast."

Master Jonas laughed. "Evelynn verified the boy fell from a shelf and hit his head hard enough to knock him out. You heard her say this with your own ears. I think he was merely demonstrating a confusion from that fall. He is recovered now; you can plainly see. You want to send this treasure away for such a temporary situation? You are the one insane, Leo, not my Maxx." He reached down and ran his fingers through the top of my wet hair.

Master Bladrick nodded. "I agree with Jonas and Claus brother. Whatever happened earlier was a response to his injury and nothing more. The boy is not psychotic, that is obvious. I go further to add that I won't approve sending him outside the Haus for medical treatment he doesn't need."

Master Leo gasped. "Bladrick, I insist you think a bit about what you say. Schizophrenia is fucking serious. This boy is under too much stress. He needs a real rest not, just a couple hours break."

Master Claus shook his head. "Mad Maxx, tell me do you feel overly stressed by your workload in our collar?"

I shook my head. "Nein, Master. I thank you for allowing me to serve all of you though I am most unworthy of such an honor." I trembled slightly terrified that Master Leo was refusing to relent his argument to have me institutionalized and trapped in my metal.

Master Claus smiled. "Well, there you have it Leo, right from the Priceless's mouth. He is thrilled to serve his Masters. I rule this concern you voice unverified. Mad Maxx will return to his rounds with each Master tomorrow. Bladrick and I will keep him two days, then he comes to serve you Leo. Jonas, you will keep him the other four days of each week since he is your blood bonded. Then you can bring him to me and Bladrick on every fifth day of your turn."

Master Jonas grinned. "A most wise and fair decision, brother Claus. I take him home right this moment if no one objects." He began to reach down to grab my leash.

Master Leo yelled out, "You wait a fucking minute, Jonas. Claus, I demand Bladrick, and you take him with you or even leave him here. I beg my brother Jonas's forgiveness, but I think he is aiding this ill collar in deception of the severity of his symptoms."

Master Jonas frowned. "Are you serious Leo? You are being paranoid. I hide nothing. I am honest about Maxx's mental illness with my Elder brothers. Why the fuck would I hide anything from any of you now?"

Master Claus shook his head. "I have waited for my turn to enjoy my full special services rights with Mad Maxx longer than I should have, thanks to that shit Drexel pulled. I will not use up my days in the cycle with him while he is still recovering from the chemical trick of the old bastard. Let Jonas take him home Leo and let this go."

Master Leo crossed his arms appearing angered. "Then I volunteer to take him in the cycle and forfeit my full special services rights. I am not such a little boy I cannot control myself from coupling with an injured collar. He stays with me, then I can watch and see if Jonas is right or I am."

Master Bladrick nodded. "Ja, okay. I agreed to this if and only if you swear no full coupling. I also demand you drop this speaking about outside medical aid when the boy provides perfect service otherwise to you. I in fact never want to hear another Gott damned word of it."

Master Jonas scoffed. "Well, I don't agree with this. Leo claims he won't try intercourse with my man in this cycle. I think that is bullshit. Once he is alone with the Priceless collar in his bed, he will break that oath no doubt."

Master Leo shot a look of indignity at Master Jonas. "How dare you accuse me of not being a man of my word Jonas. I say I won't fuck this Mad Maxx round and I won't. If you cannot trust me, then why do you lie and call me your brother."

Master Claus raised his hand "Enough, I agree with Bladrick and Leo on this one Jonas. Truth is you are just as likely as any of us to breach the boy's healing areas. I think you intended to do that very thing tomorrow night before turning him over to me. I know you, brother. I don't mean to insult you, because I myself would be more than a little tempted with Maxx laying in my bed and no one looking. If Leo says he can keep his lust under control, and this will settle his nerves, then I see no harm in it. You are out voted Jonas. Mad Maxx stays here with Leo and tomorrow night at midnight the boy's collar is turned over to me and Bladrick. This is my ruling, and it is final."

I nearly opened my mouth in insolence at that bullshit. I didn't want to stay with a man that said I was crazy and never getting out of my collar. I never hated Master Leo more than I did that moment. I began to think he could join Xavier, Barnim and Drexel. This bastard was gonna be worm food.

Master Jonas growled in anger. "If this is the ruling of you all, I will respect it. However, the boy is without clothing nor his supplies. I will take him home to pack for his rounds unless there is a fucking issue with that too. Perhaps all you predators prefer him to stay naked? His being dressed would do nothing but slow your lustful attentions down. After all, you're all in such a hurry to dip into the forbidden silver you cannot even wait to let the poor boy heal the hell up."

Master Claus rolled his eyes at that. "You need not insult us, Jonas. Ja, take this collar to your apartment and have him collect all that he will need to be away from home for three full days. I think you will find he is returned to your bed no worse for the wear."

Master Jonas reached down and snatched my leash with vigor. "Don't make promises you don't intend to keep Claus. Come on, Christian Axel." He pulled me behind him storming from Master Leo's apartment with me wearing nothing but that towel.

We arrived at my Master's home rapidly. He practically kicked in his front door he was so pissed over Claus's ruling. I was a bit more than a little frightened. I worried the Vampire would take out his anger on my flesh.

Once inside he turned to me and spit out with much irritation. "Go get dressed. Make sure to wear heavy clothes and a fucking coat too. You are gonna need it the second the squid fingered Leo is alone with you. Better pack all your supplies. Those foul old men are going to fuck you till their dicks fall off. I won't stand for you to return to me tainted by their lust. I hate sharing my Priceless with these idiots. I hate every fucking one of the cocksuckers, especially that rat Leo. Just thinking of him touching my beautiful Christian Axel makes me sick." He threw his keys across the room in fury.

I ducked then ran quickly to the room with the grey door. He hated thinking of them touching me. Shit, he should have been the poor fucker they were molesting. I said nothing of that though. I rushed to my closet and grabbed a complex outfit. I pulled my clothing anxious that Master Jonas would come in and see the terrible marks left by Mistress Heidi's thudder.

I swear I never dressed in so many layers faster than I did that afternoon. I even managed to get a full makeup job done within thirty minutes. I grabbed my canvas traveling bag and packed it full of fresh clothing, my makeup case, and all my hygiene supplies. I grabbed the jacket with my Annette hidden in the pocket and returned to my still fuming Master Jonas sitting on his couch. He looked up and saw me as I fell to my knee waiting for his commands.

He snorted. "You are getting good at this makeup illusion. I cannot even see the bruises anymore." He reached out and pulled up my face to examine it more closely.

I winced as he poked at my swollen left cheek. "Master, I apologize for causing all this trouble with my moment of brain weakness."

The Vampire groaned as he stopped probing my facial marks. "There is no need for your apology, my love. Leo is the nosey bastard. Always has been. The fucker is the son of one of the wealthiest families in all of Germany. He is the rich bored, Christian Axel. All he does since I ever known him is travel round the Haus looking for drama to watch or get involved in. I don't believe for a second, he won't demand your special services in full this very night. I will give you the numbing cream for this injustice. It won't stop the damage his couple will cause, but it will keep your pain to a minimum. Make sure to apply a thin layer to your tissues at least ten minutes before he enters you or its effects won't save you from the terror of his intercourse. You can use it up to three times a day but no more than that. It is powerful stuff, and you can overdose with it." He reached onto his coffee table and handed me a small blue jar.

I took the container looking at it in confusion. "You mean you had this all the time, Master? You could have given me this sooner, I say that with respect." I was not ugly in my tone but to be honest I was pretty pissed off he let me suffer when this stuff existed the whole fucking time.

Master Jonas shook his head. "Nein, I only managed to get this anesthetic this morning. It took some time to talk the Haus doctor into giving it to me. That greedy bastard told me he needed all that he had and wouldn't sell me some for any price. I finally told him it was for a hemorrhoid that been bothering me and not for my collar's comfort during my intercourse with him. That sonofabitch wanted me to prove I had one. Can you believe that? Well, I was foul enough in my threats he dropped trying to get proof. He didn't trust me a bit. That pissed me off to no end but there it is. I got it at last." He chuckled bitterly.

I nodded. "I thank you for the mercy of this Master. It will help more than you can imagine." I packed that jar of salvation into my bag with a sigh of relief.

My Master frowned "This will be the only time I can get my hands on this stuff. Use it wisely, Christian Axel. We will have to return to Leo's place this minute but first I want your adoration for the trouble I went through to get you some mercy from the pain of being the bottom of the couple with your Masters."

I shot a look of fear. "Adoration, Master? Do we have time for that? I mean I thought you said I need ten minutes for the anastatic to work." He interrupted me with a loud laugh.

"Calm down, Christian Axel. I am not demanding to fuck you. I meant where is my kiss and cuddle for all the trouble I went through for my man." He grabbed my bat collar and pulled me into his lap for deep kissing with his heavy pawing of the boy for several minutes.

I endured his affections and provided perfect service back to him. The Vampire had much difficulty pulling himself from his growing interest in this lustful embracing.

He finally groaned then pushed me off his lap with a look of regret. "This is bullshit. The day I get the substance that would allow me to enjoy my treasure once more is the day Leo pulls you out of my hold."

I dropped my gaze at that saying nothing. For the record, it made me more than a little angry that he was indeed planning to force his couple despite knowing I was far from healed from that chemical burn. Granting pain killer to keep me from suffering his thrust would not stop the injury it would cause. Of course, that rapist was neither providing any pain killers nor respecting my damaged tissues.

His complaints that day were a loud reminder that Master Jonas didn't really love me as he says he does. If he did, he would have respected me and allowed me the full time to recover properly rather than just offering a bit of mercy while he forced himself on me. I stood up and followed behind him in silence feeling damned low that this was the life I had found for myself.

I listened to him grumbling nonstop as he led me back to Master Leo's apartment. He knocked only twice. Master Leo came to the door smiling, trying to appear friendly.

Master Jonas scowled. "Here is the Priceless, Leo. You better remember the oath you made. I will be asking him if you broke your word. He was told I would beat him if he dared to lie to me."

Master Leo nodded. "You have nothing to worry about Jonas. I won't lay a lustful finger on him until it is his time to be transferred to Claus and Bladrick. I keep my word; this you can trust." He held out his hand waiting for the Vampire to hand my leash to him.

Master Jonas slammed it into his hand. "I am not worried about it, Leo. Mad Maxx can hold his own with the likes of you. Good luck not falling from the top of stairs or banisters without wings, brother. I wouldn't want to be you. He likely has been most offended at learning you wish to see him locked in that collar of mine for all time. Buh-bye, Leo." He laughed evilly and tore off down the hallway leaving me staring at the floor in the claws of the notorious schwuler Leo.

Master Leo cleared his throat nervously. "Uhm, come inside, Maxx. No reason to give drama to the entire Haus, ja?"

I walked in and he closed the door behind me. I didn't even look up at the bastard. He stood there, appearing unable to decide what to say to me. Then he gained his tongue.

"So, you are likely hungry being the growing boy, ja? Would you like me to call in dinner for us?" He looked me up and down.

I shrugged. "As you wish, Master, if you are in need of a meal, but I am not hungry thank you for the generous offer."

He frowned at that. "Not hungry? You are already too thin, Maxx. This lack of appetite is common with your sickness. My own sister was like the rail. The girl never eats more than a morsel at a time."

I shrugged again, "Then I don't disappoint you since you seem to know everything about me, Master."

Master Leo sighed. "I don't wish it to be this way between us, Maxx. I didn't mean to upset you earlier. I merely worry about you is all."

I scoffed. "Why would that be, Master? Even if I had this disease, you all seem to think I have, then I would still be capable of sucking your cock on command, wouldn't I? If you desire my special services, then give the command. Otherwise, I have studies to attend for my schoolwork.

Thank you in advance for the mercy of getting the bona fide business of your demanding my leash tonight out of the way."

My Master appeared startled with a sudden flinch. "Nein, I don't request special services, Maxx. I am a man of my word. I demanded my leash rights to give you a rest from lustful touching and I stress for no other reason. Go study on the floor over there where I can keep my eyes on you. Get to it if this is something you need to do." He pointed at the peach-colored leather couch. Thankfully, he got rid of that weird orange cube thing of Drexel's.

I looked at him with surprise. "You aren't going to force your couple, Master?"

Master Leo rolled his eyes. "Why the fuck do all of you seem to think I am the sex fiend? I already told you, I keep my hands to myself. Go study damn it, or do you wish to insult me some more?"

I shook my head. "Nein, Master. I beg your forgiveness for my errored perception. Where shall I put my things?" I held up my overnight bag.

He pointed down the hallway. "In my bedroom by the foot of my bed. Then return and do this schoolwork. I will read quietly while you attend your studies."

I did as I was told and got out my books. I sat down at his feet without his attempting to molest me in any way. I then spent many hours catching up my education. I was way behind thanks to those eight months of trouble and the struggle to serve the brutal Elder Masters. To my surprise Master Leo never interrupted nor requested a single service the entire afternoon nor night.

He ordered food and had the trays brought up by the black collar courier for the Elder's floor. He laid the plate next to me, but I wasn't hungry. I thought it best not to bother eating anything the Masters brought to me anyway. Master Jonas apparently had drugged me then paid a fake doctor to say I was a schizophrenic. I wasn't falling for anymore tricks of these Dominants.

Master Leo seemed hell bent to see me trapped in my metal. It would make sense he would drop something in my food or drink that would make me appear crazy. Well, he could forget that shit. Mad Maxx was no one's fool. I politely told him I would eat it later. I waited till he fell asleep over his book. With much stealth I threw that tainted crap into the trash can. I decided from the second on to avoid all food or water other than from the Haus taps.

It was now clear to me Master Jonas had been using me in his plot to take over the power of the place. I could no longer ignore the fact that every fucking Dominant in there, and a few black or even silvers, were out to abuse my abilities for their own conspiracies. I returned to my place on the floor and watched Master Leo snoozing with great suspiciousness.

Christian, Mad Max and Max all showed up, rested at last I suppose. I shot them a look of fear.

Christian snickered. "He looks like the Aardvark brother. Only Helga and Heidi are uglier."

Mad Max scoffed. "Correction, Christian. Claus is ugly too, brother."

Christian nodded. "Ja. He is. Okay, Maximillian good job finding out about the drugging. I admit I was fooled into thinking we were sick. To think Master Jonas did all that just to get Peter to toss our collar. What a cocksucker he is. Eight months of acid trips. That is low."

I sneered. "Perhaps, but then again you can say that can't you, Christian? Wasn't you having to be fucked by our father, was it?"

Mad Max and Mad Maxx laughed loud at that, then Mad Max said, "He got you there Christian. I for one would suffer eight months of a bad trip in the dungeon to avoid sucking Peter's cock. I hate to agree with Maximillian, but he has a point."

Christian shrugged. "Fair enough but beware. If the Vampire did this once to steal us from Peter, then what may he do to stop the leashes of these foul Elders."

We all looked at each other with fear in our heart. Christian had a point. Mad Maxx and I had heard the furious jealousy in the Vampire's voice over having to hand us over to his brothers. We all agreed no matter what shard was running the boy nor how hungry the flesh got, no one allowed any food to enter us.

Then Mad Max suggested we should stop taking all the pills Master Jonas was giving us as well. Medication for the schizophrenia is what he called it. We all believed it more likely was poison that kept us confused so he could mislead us. We unanimously voted, no food, drink, pills, or shots permitted. Let's see that Vampire mess with our thinking without his mind controlling substances. We congratulated each other on solving this mystery of that horrible delusional world that kept pulling us under into madness.

All that night we quietly spoke about our future targets, plans, and designs for our future as the Dominant. Many laughs and some arguments were shared among us. Master Leo slept soundly on his couch with his book in his lap during the whole meeting of all the Christian Axel shards.

When at last the clock struck nine the next morning Mad Max, Max, and Christian took the wheel. Mad Maxx and Maximillian took to our slumber immediately to prepare for the rounds with our Masters that would start at midnight.

Our evil team could handle the boy until sundown. Besides the work that day required the Sadist and Killer of our mind. The time had come for the end of the H twins down below. We had waited many years to end those Harpies reign of terror and killing of little children. The evil bitches time had come to an end.

I got up and stretched, then looked at Christian. "You ready to do some equal service return, brother?" I giggled most wickedly at that.

Christian nodded then shot a look at Max still melted to my leash. "Sure am, brother, but I have to ask you, is this bastard going to stand their smiling like that for the whole thing?"

I growled with irritation at his pointing out this moron to me. I was trying to ignore that fucker. "Don't pay him any mind. If we pretend he is not here nor ever speak to him maybe he will go the fuck away."

Christian groaned out. "I will do my best to do that but fuck Mad Max that goofy grin gets on my last nerve."

I raised my brow in confusion. "hat? Brother, I don't mean to be rude, okay ja, I do mean to be rude. This motherfucker used to hold your collar, didn't he? I would think you're used to putting up with his shit by now."

Christian's eyes went wide in shock. "Huh? Nein, this shard is not the shard I knew. This Max is different, brother. Whatever the hell he is, I don't like it. I wish you would call Der Hund and have him exorcised or whatever the fuck our Core does to these malformations. This Max makes me feel weak when he comes around me. He scares the fuck out of me, too. So, there it is. I said it. You get rid of him Mad Max. I mean it or we will have to stop working together soon. This shard is doing something to me. Max is making me, I don't know, change somehow."

I shook my head snorting at that bullshit right there. "Stop being a Pussy, Christian. This stupid Max shard is a nothing. He doesn't frighten me a bit. If he gets in my way, I will beat him with my cane. You stand up to him and he will run like the coward he is."

Christian grumbled. "I don't know how you cannot see this horror attached to your leash Mad Max. I tell you this thing is dangerous to me. He wants to see me shattered. You can see it in his eyes. See there, he is smiling about my fear of him. You get rid of him, or I won't stay in the flesh with you ever again no matter what Der Hund says. I mean this." I watched in shock as Christian moved as far as possible from Max and me trembling in terror over this nothing shard.

I clicked my tongue and rolled my eyes at this weirdness. I didn't have time for it. I quietly knelt at Master Leo's feet and called out.

"Master, I respectfully request your release me to go to breakfast and my lesson with Egon. If you would be so merciful to allow this, I shall return by eleven." I said in a near whisper.

Master Leo sat up rubbing his eyes and groaning. "If this is your normal schedule, ja, I release you. Be back here by eleven no excuses."

I nodded. "Shall I bring you back a breakfast tray, Master?"

He smiled with kindness. "Nein, I will get my own breakfast in the Great Hall in a bit. If you need nothing else, then get going. I'll see you in a bit."

I started to get up but then stopped myself. "Uhm, Master. Could you loan me a pair of sunglasses? I thank you in advance for this mercy."

Master Leo gasped in surprise at my odd request, "Sunglasses? Mad Maxx what the fuck would you need those for? There is no need for eye protection from the sun in the fucking hallway or torture chamber."

I nodded. "Ja, you are correct, Master, but both my eyes are black. I don't like the way everyone stares. If I covered my face with them no one sees the injuries."

Master Leo chuckled. "Ah, okay you worry about everyone thinking you're insolent. I forget that can hurt a Priceless's reputation. Ja, I go get you a pair, but you bring them back in one piece I mean it. They are not cheap ones." He got up then returned from his room with a large pair of fashionable high-quality shades.

He put them on my face and made many compliments that I "looked handsome in them." I didn't give a shit about my looks in those glasses. I just wanted to protect my eyes from anyone spraying shit at me. I got up and without further hesitation rushed from his apartment and for the stairs. I would have to run the whole way if I didn't want to be late for my date with my Mistress lovers. *You see Meine Liebe, I planned on them being the late ones. Mad Max winked at me with a wicked chuckle. I giggled at his clever play on words.*

I raced down the hallways reaching the door of the Dungeon stairwell without a moment to spare. I removed the glasses dropping them into my hidden pocket. I took a deep breath then I started down the steps. I was greeted by Mistress Heidi before I got even one third of the way down.

She was wearing the most hideous green and black flowery knee dress. Her neckline plunged to reveal her wizened bosom. I saw she was not wearing nylons and her legs were hairier than my own. I didn't think she could turn me off more, but that Billy goat managed it.

She smiled at me with a look of lust. "What do you think Mad Maxx? Like it?"

I gulped back my bile and forced a smile of thrill. "Ah, ja. This is, uhm, perfect Mistress. You are truly a vision (of what I was unsure). I am not worthy of such beauty. Maybe I should turn around and leave you for the Adonis you deserve." I forced every one of those words, let me tell you.

Mistress Heidi whined out. "Nein, don't you their leave me, Mad Maxx. I adore you. I need you. Come to me now and give me the affection you swore I deserved. I command it."

I smiled as the demons of my illness rose within. "Your wish is my pleasure, Mistress." I stepped down while she came rushing up.

We collided on the fourth step down. The wanton Mistress pulled me into her eager embrace grabbing the sides of my head by the hair. She forced my lips to her own pushing me into the stone support wall. I whimpered when she began to pull at my blouse and grope my crotch roughly. I tried to pull away from her tentacle arms, but the woman would not relent.

Let go of my man this second, you bitch. Mad Maxx is mine, you whore," Mistress Helga shouted out from the bottom of the stairs.

Mistress Heidi pulled away in a startle. "What? Helga? Fuck off. He belongs to me. You better mind your fucking business."

Mistress Helga scoffed. "He doesn't want your tired old ass, Heidi. This boy loves his Mistress Helga. You are trying to steal him from me. You have always been jealous of everything that I ever had. This time I will not stand aside and let you get away with it. You come down here and face me, you bitch."

Mistress Heidi snatched my leash into her claw and dragged me behind her down the stairs. The two Dungeon Mistress's faced off. I noticed that Helga wore an exact copy of the hideous dress her sister selected. I watched in slight humor as the two women glared at each other for a moment sizing each other up.

"Well, here I am sister. You were sounding brave when I was up the stairs. Now your lion mouth is quiet like the mouse?" Mistress Helga looked at me appearing frightened of her angry twin, who was the meaner by reputation.

I opened my eyes wide as if terrified, then silently said, letting her read my lips, "I thought this was you, my love. Help me, Mistress." She nodded that she understood.

Mistress Heidi growled out. "Again, I say, sister, leave me the fuck alone. Mad Maxx is mine."

Mistress Helga looked at the floor. "Ja, ja, I heard you sister. I leave you to it. Forgive my intrusion." I almost died when she said that. I was sure to be torn apart by Mistress Heidi.

Mistress Heidi chuckled. "That is what I thought. Get the fuck out of my face sister. Go off and look at your pictures of the pretty boys in the magazines. You leave the real affairs of the heart to me. This Priceless is too fucking good for you anyway. Tell you what, I let you kill that bothersome silver you hate so much later when I am done getting my taste of the forbidden silver. If you are good, I will even tell you about my exploits with him. How'd that be?"

Mistress Helga nodded looking as if she may cry. "That sounds fair sister. I will go now and leave you to it. Apologies for my harsh words." I watched in terror as the woman walked away leaving me in the cruel clutches of her twin Mistress Heidi.

The brutal Mistress turned around smiling in victory. "Sorry for that regrettable interruption, my love. Now where were we? Ah, I was enjoying my tasting you.' She come back at me once again slamming me into the wall forcing her kissing and molesting my flesh everywhere.

I held my breath doing my best to endure this foul situation. I sent out a cry for Maximillian. I am not designed for this seduction shit. I mean if the woman was good looking, ja, I can handle that, but this thing looked like a man or a farm animal. Not the cute ones either.

Several moments went by without an answer from that cocksucker Maximillian nor a pause in the rough handling by Mistress Heidi. All of a sudden, the Mistress let out a loud gasp. She pulled her face away from me. I was startled to see her eyes were wide, and a small trickle of blood was rolling down the corner of her mouth. I whimpered when the woman staggered backward then turned around.

I saw the blood pouring out of a wound in the center of her back. She had been stabbed deeply. Mistress Heidi then let out a gurgling yell and fell onto me. I groaned, pinned to the wall by her weight. I felt her tremble just as her knees buckled. The Mistress slide down slowly then fell to her side, her eyes were open and fixed. She released her final breath then was silent and still.

I stared in disbelief at the growing puddle of blood that was coming from her breast. She had been punctured in the chest when she turned to face her attacker. I glanced up in terror to see Mistress Helga holding a huge kitchen knife. On the blade of it was crimson fluid. Mistress Heidi's twin sister had murdered her over me.

Mistress Helga sniffed back tears while gazing at her dead sister. "I am sorry Heidi, but I am sick to death of having to watch you take everything I ever wanted away from me. Now, you realize I was the better of us." She then looked at me with a look of despair.

I shook my head. "My love. You killed her. Oh, my Gott. We have to hide this crime, or they will send you to prison." I was glad that for a change I didn't have to be the one to murder a target.

I couldn't believe my good luck. Helga just did my job of stopping both of them from ascension to Elder for me. I could use my knowledge of her ending her sister to make her leave the Haus forever. The children in silver would be saved from her and her sister's murderous hands. Best of all, I didn't have to stain my soul with another killing. This was a happy moment for old Mad Max, then I noticed the look in her eyes.

She apparently was thinking the same thing as I was. Mistress Helga raised the knife above her head. She slowly began to walk towards me.

I gasped then gulped. "Mistress? My love, put down the weapon. Don't do this. I won't tell anyone. I said I will help you escape the law. Why would you want to kill your lover, Mad Maxx? This violence is not necessary." I looked to the steps ready to try to outrun this crazed woman.

Mistress Heidi shook her head. "I must kill you, Mad Maxx. You will use this to control me, and I cannot have that. Plus, my sister has tainted you with her touch. I will not tolerate her hand-me-downs any longer. You must die." She let out a shriek and came at me like a leopard.

I rushed for the stairs as fast as my legs could carry me. She was faster than I could have imagined. I felt her reach out and grab my arm just as my foot hit the first step. I let out a loud yell for help as she threw me to the floor with great force. I rolled across it knocked nearly out of my senses by her insanity driven strength.

I got up to my knees trying to escape the Mistress already nearly within striking distance. The woman stopped her attack suddenly then let out blood curdling scream. I cowered in terror as the woman come tumbling down on top of me. I yelled out in fear thinking this was the end for Mad Max. I groaned in agony as her heavy flesh crashed down once again pinning me to my spot. I could hear her gurgling into my ear as I screamed out in terror of being underneath her like that.

It took a few moments for me to struggle out from under and away from the now quiet Mistress thanks to her incredible weight. I then noticed she wasn't attempting to restrain me from escaping her. Once I was free, I crawled on my hands and knees till I reached the safety of the wall on the other side of her. I never took my eyes off the Mistress, but she never moved even a muscle.

It was then I noticed she had no weapon in her hands. She didn't appear to be breathing. I gasped when a crimson flood started spreading across the dungeon rock floor from around her head. I heard a noise by the stairs. With a startle

I glanced up to see Master Leo standing their holding Mistress Helga's knife in his hands. His hair was disheveled from a recent battle. I saw the blade of the sharp he held was covered in fresh blood.

I realized at once that Master Leo had slit Mistress Helga's throat. She was as dead as her sister.

Chapter 43: The Strubenfarm

I couldn't tear my eyes away from that bloody knife Master Leo held in his hand. He was panting and agitated. I blinked several times unsure if I was seeing this scene correctly. The rat bastard, betraying, wimpy Leo had murdered the raging Mistress Helga to save my life? Couldn't be. I must have been drugged again.

Master Leo trembled. "Maxx, come to me, love. We have to get the fuck out of here before someone sees us. This is not easily explainable." He motioned for me to head his way.

I shook my head shaking in terror of my own. "I demand punishment, Master. I won't come. You killed Mistress Helga and hold the blade you done it width. I will warn you I can fight. I am not an easy kill. Be ready to find your grave if you intend to finish me off." I stood up slowly doubling up my fists trying to look brave. I admit I was fucking scared as hell. This guy was bigger and held a weapon. I had no chance, you know.

My Master looked down at his hand holding the sharp. "Oh, meine Gott. You think I intend to kill you? Nein, why would I do that? I just killed Helga to save you Maxx. Look, if we run right this minute and I plant this knife in Helga's hand it will look like a murder suicide, ja? Everyone knows the girls hated each other. No one needs to be the wiser. I will have to trust you to keep your tongue still, but I know you will." He rushed over and put that blade in Mistress Helga's dead hand then motioned at me to come to him again.

I took the stance of aggression. "I take the punishment, Master. I said this already."

Master Leo looked with fear at the door of the stairwell above us. "We don't have time for this, Maxx. Punishment denied." He rushed me and grabbed my leash tearing off down the hallway dragging me behind him before I could get more than one single punch into his chest.

He ignored my blow seeming to not even feel it. I whimpered and grabbed my leash trying to pull it away from this madman killer. He held tightly and was much stronger than he looked. I was forced to run behind him or be pulled to the floor and dragged.

He didn't look back nor say a word about my trying to resist his hold. The man was hell bent to take me wherever he was going. I realized in great fear that Master Leo wasn't kidding, he intended to flee this murder spectacle. I was coming with him like it or not.

My Master didn't run us back up the stairs, which is what I would stupidly have done. He knew someone would spot us coming out of there without a doubt. Then later, when the girls were found, we would be the suspects. With that in mind, he hauled me down the hallways of the Dungeon taking us deeper than I had ever been in all my days while suffering in that place. He kept to all the shadows keeping us from being detected by anyone that passed by. This man seemed to know every secret nook and cranny of that entire rock prison.

I had given up trying to break his hold on my leash. I ran behind the panicking Dominant hoping he wasn't taking me to some dark corner to wait a minute, Master Leo was over six foot tall. He was fit, strong as hell, and without facial hair of any kind. The man had been trying to fuck me for over two years but other than a blow job had been unsuccessful. He always seemed to show up wherever I was when drama or a rape happened. He was obviously following me and that is how he arrived to pull Mistress Helga off me.

I gasped in pure horror at my sorry situation as I thought, "Oh shit, the rapist is my Master Leo." That sonofabitch.

It was clear to me I needed to kill this rat bastard. Truth is, I was happy to murder this target, more than you can know. I had hated the man for almost two years since the day he roughly probed me to find my anal virginity gone. He had been involved with my father and mother in their plot to put that hideous Cora on the throne. He had even forced a leash with Ben (Mistress Cora was his Dominant, remember) which led the boy to hate me like he did.

As I solved the mystery of my attacker, I wished I had grabbed that fucking kitchen knife. I would have no choice but to kill Master Leo with my bare hands, thanks to leaving that weapon behind stupidly. One thing was for damned sure, he wasn't dragging me into dark corners or closets to violently fuck then beat me anymore. I shot a look at Christian.

"You heard my thoughts, brother? This is the rapist. Kill him. I don't fucking care how. Just make this his last they on this Earth," I growled out at the murderous shard.

Christian nodded his eyes glowing red with hell fires. "My pleasure, brother. You and that monster get away from the wheel and I make this happen."

I smiled with evil and pulled back hauling the goofy Max shard with me. Christian took the boy's mind. It was time for Master Leo to face the wrath of his victim for a fucking change. At that moment, the rapist Master pulled us into a small dark rock room that had apparently not been used in many years. Dust was everywhere and the dankness of the hole was near smothering to breathe in.

Master Leo backed into a wall completely out of breath from the racing. He put his arms on his knees and leaned forward trying to pant himself to vigor once more. I watched Christian glancing at me with a huge toothy smile. He was about to show me his talents at sending out enemies to their grave.

Master Leo didn't notice Christian creeping up on him with stealth. "I think we are safe for a moment. We wait here a bit, calm down the sweating, then we go back out and stroll along as if nothing is wrong. Either someone finds the girls in their death state, or we come along and raise the alarm ourselves. We make sure everyone possible sees us when we leave this hidden spot. No one will put us at the scene of this horror if we play it cool and calm. If worse comes to worse, I will confess to the murder to keep them from coming for you Maxx. I will tell them I was having an affair with the girls, and you come along to serve me after the death was done. I swear it on my life, I would die before I let them hurt you anymore than they already had, my love." He gulped in his air watching out the sliver of the door his back to the stalking Christian.

I heard his words but had difficulty understanding them. He said what? That he would take the punishment rather than let the Guard come for me? That could not be right. This had to be another trick. A mind game. Master Leo was a worm, a scum bag, a perverted schwuler that preyed on the innocent. He was also the rapist, wait, why would he rape me?

He didn't have to attack and beat me. He is my Master and can command I do whatever he wanted. He had been alone with me all night and never laid a finger or anything on me. Nein, this was a trick. Master Leo is just clever, that is all.

Christian was about to wrap the boy's hands around Master Leo's throat when all at once Max flew into a rage. He tore forward with the ferocity of a wild animal dragging me helplessly behind him. He pushed Christian away from the wheel knocking the shard to his back. He began to kick him with all his might.

Christian howled and wailed in pain and terror as Max beat the shit out of him with his feet. I rushed to grab Max and pull him off our brother, but I was knocked backward. It was like there was a forcefield around Max. I flew at him repeatedly but was forced back, unable to even get close to him.

To my utter terror I noticed the ground beneath our feet was trembling with a massive quake. I looked out the boy's eyes to see lights flashing, then suddenly the flesh fell to its back. It went into a series of heavy spasms of a Grand Mal seizure. I watched in disbelief as every time that Max landed a blow on Christian the boy would flop around. Their fighting was causing this brain misfire.

I screamed at the top of my lungs, "Brothers, cease this battle. You are killing the boy. Gott help us. Max, let Christian go."

Max turned around with the look of a demon in his expression. "Call of this killing of one that loves us, fool. I will kill us all before I allow this mad killer to murder the innocent. Do it Mad Max or I destroy the boy. I mean it."

I panicked at his words and his look of the devil in our own soul. "Okay. Christian, I take the wheel brother. Leave Master Leo be. I withdraw the order to destroy him. He is off limits."

Christian wailed out, "You take the fucking wheel. I am out of here. I told you he was trying to shatter me. I vow to never travel in the flesh with you again. I curse you both to hell." He rolled out of Max's foot reach and fled the flesh running fast as possible hiding within the shadows of that place.

I was beyond terrified. Der Hund said Christian was to be my partner in smiting out enemies, but Max just fucked up the program over the fucking pervert schwuler Leo. Really? Even if he wasn't the rapist, he was disgusting and a child molester. He was a lot of terrible things, but innocent? Max was not only a pain in my ass, but he was also fucking insane.

I took the wheel sweating and panting in my fright at the likely shattering Der Hund was going to impose on me. I felt the ground stop trembling almost immediately. I shot a look of hate at Max.

"What the hell have you done, Max. You and I are toast the second Christian tells Der Hund of this bullshit. Fool, you killed us both. Why?" I kicked the wheel and the boy spasmed another second from it.

Max suddenly lost his look of demonic influence. He was calm, serene, and at peace with that Gott damned goofy grin once more.

He chuckled with thrill. "Der Hund ordered me to attack Christian brother. Master Leo is our ally, not our enemy. There will be no punishment for minding our Master Core."

I nearly fell to the floor at that madness. "Holy shit. You are crazy as hell, Max. I didn't see Der Hund anywhere. You say you were ordered by him. Are you hearing voices, are you?"

Max laughed hard at that. "Mad Max you are beyond dense. I am the Soul fool. I know Der Hund's deepest thoughts, desires, and dreams. I am him and he is me. In a short time, you will see the wisdom in what I just did. You and the other shards can only know what you perceive as you run the wheel. This makes you all blind to truth and reality. You think you only saved Annette for your intercourse, but that was a lie. You love the girl, and she loves us back. Gaze upon another that loves without boundaries and total selflessness. Master Leo would give up his life in this Haus for our safety. He is willing to lose us to see us treated in the hospital. He risks his own life to save ours from that Helga that was going to kill us. He followed because he is concerned for our safety. He was protecting not stalking. Wake up Mad Max. Leo loves us for truth, unlike any other in this Haus. Even Annette cannot boast such a pure feeling for the boy. When love is real you are willing to give up all you are, own, or ever will be just to see the lover get the best of everything. This is Leo for the Mad Maxx."

I stood there staring at that dumbass for several minutes not sure if I should laugh or be pissed. This motherfucker was beyond bat shit crazy. I finally decided to just ignore him and await my fate with Der Hund when that Core of ours kicked my ass for losing my brother Christian. I redirected my attention to the ailing boy. He was awakening from his nasty seizure, thank Gott. I thought for sure we killed him.

I opened my eyes to find Master Leo kneeling next to me holding his quirt handle in my mouth, like a bridle for the horse, ja. I was feeling groggy, woozy, and confused. My Master opened my eyes one at a time, looking into them best he could in the dark room. He shook his head and maintained his pressure on that thudder in my mouth.

I groaned out and tried to move. "Hold still Maxx. You had a nasty seizure, love. Can you hear me? If so nod but do it slowly. I need to check you for broken bones. The spasms were quite violent." My Master watched my head closely.

I nodded with gentle movement as commanded. He gasped and smiled with joy, then removed the quirt. He reached out and felt my neck, then ran his hands along my flesh. He stopped at every joint, moved it gently asking if that caused pain. I shook my head with each bending spot.

When he finished his examination, he sighed then sat back to his bottom. "I thought you were a goner. love. You cannot stop taking your medication Maxx. I noticed you didn't take any last night nor this morning. I bet if I spoke to Jonas, he would tell me you missed doses while with him too. Suddenly ending the meds will cause seizures like that."

I watched him wipe the sweat from his forehead while staring at me with sternness in his expression. "I took it Master. You are wrong in thinking me missing my medications."

He looked shocked. "Maxx, you dare to lie to me. I know Gott damned well you are not taking the medicine, boy. I counted though pills when you arrived. I noticed after you left, the same amount was there this morning as last night. Look don't bother to deny it. My sister pulled this shit of believing we were trying to control her mind or poison her with the doctor's medication. You are paranoid, hallucinating, delusional, and my love, in a lot of danger of death from this bullshit. The others may not fucking care, but I do."

I frowned, not trying to move yet. "Master Jonas said he paid that doctor to lie to Peter Master. He says he loves me and cares, but he got that man to lie. You tell me not to lie to you, but all you Dominants do is offer tales to me. You are wrong about my having schizophrenia. My man is slipping me mind altering drugs, though I don't know why I confess."

Master Leo scoffed. "My love, there are no drugs out there that cause the symptoms like you show them to me in only the last two days. Jonas paid that doctor to fib about the severity of your disease, not to say you have something you don't. I can tell you why he did it too. Around this Haus most only care about their fucking hard dicks and power plays no matter who the hell gets crushed by it. I am sick it. There is more to life then fucking and being admired. Not a Gott damned one of them ever stops to think of the pain they cause others in their big rush to impress the nothings that live in these walls. They hold a precious treasure in their hands, but even that is not enough for any of them. How the fuck can anyone find happiness when there isn't better than perfection, but even that doesn't impress any of them."

I winced at his brutal but truthful words. "I am doomed aren't I, Master? I will end up in the yard like all the other Priceless. You can be honest with me. I am not a kid. Not for a long time now thanks to Peter and others." I shuddered holding back my sudden despair over that admission out loud.

Master Leo looked at the floor appearing sad. "I cannot answer for what others do. I can only apologize for my part in it all before I grew the fuck up only a year ago when I saw Peter toss the collar of the most perfect boy I ever have seen. I have damned myself a million times for falling into the same moronic mindset all of them have. I hurt you my love and I can never make that up to you. I can only say I will never do it again long as I live. You asked me if you are doomed? Nein. Don't you worry. Leo is here for you. I swear I will do everything in my power to protect you from the evil designs of those who abuse you to reach their dumbass goals. I couldn't save my sister from the demons that haunted her. However, I am going to find a way to save you, my beloved Christian Axel. Even if I end up getting thrown out of the Haus, so be it. I do this knowing you will never feel anything for me in return. Peter told me you are straight. I respect that and I do apologize for that scene in the torture room in front of Peter. I will never force you to sleep with me or force my touching on you again, that I also vow to you. I don't need to sleep with you to find my satisfaction."

Ma eyes went with at that last statement. "Huh? You say you will never ask for special services from me, Master? Am I dreaming this? It is a trick, ja? How the hell can you find your satisfaction with me otherwise?" I admit I was curious to hear this line of horse shit for sure.

Master Leo chuckled. "Love is not sex, Christian Axel. The act of intercourse is meant to be an expression of it, but it is not the true emotion of it. I respect you, my love. I never raped anyone in my life. Ordering you to my couple is the same as forcing. I never took a collar to prevent that shit. I instead enjoyed the leashes of those that were willing. If they said nein to my advances, then so be it, another would say ja. You have grown into a gorgeous young man of amazing inner strength. Cupid's arrow has fatally pierced me, and I am helpless at your feet. Nothing else in the world matters to me anymore but keeping you as comfortable as possible. That will always be enough for Leo." He reached out and stroked my cheek with that expression of adoration unlike any I had ever seen before.

I scoffed. "Did Ben say ja, Master? You raped him. He couldn't stand the schwuler sex you had with him. He said so."

My Master furrowed his brow. "Ben? That big brute silver of Cora's?" I nodded.

He blew out his breath then looked at the floor. "Christian Axel, I never slept with Ben. I never had a leash, nor even knew of him other than of Cora's using him to bully you."

I snorted. "Why should I believe you, Master? You admit to being involved in a plot that got me hurt badly. Now, you say to trust you. I cannot think of a single reason I should. I say that with respect.

He nodded. "I cannot either, Christian Axel. I didn't ask you to trust me. I wouldn't buy a fucking word I told me if I were you. That doesn't matter to me. I say to you that in my home you will find no forced special services. I tell you I will keep my vow to aid you in your fight for your survival among these many pigs. Your faith in me isn't required for any of that is it? I love you with my heart and soul despite knowing you don't love me back. I get what I deserve for what I did, but that doesn't change my feelings for you. My soul is light as a feather just confessing this all to you. I will do you the honor of never bringing any of it up after this minute. I thank you for the mercy of listening to this old fool and his stupid dreams. The one you find you heart with, she will be the luckiest person on this Earth. You can tell her Leo is jealous of her. Alright enough of this gossiping about the fluff, ja? Do you think you're strong enough to walk?" I nodded as he laughed then offered his hand to me.

I couldn't believe this. Max was right. Master Leo was in love with us for real assuming he was being truthful. He was vowing to be our ally just as that shard told us. I shot a look at that crazy bastard. He crossed he arms; smiled, then nodded his head.

"Oh, fuck off Max. You don't know everything. That was just a lucky guess. It's all bullshit anyway. Bold faced lies; I tell you." I mumbled under my breath in irritation as Master Leo helped me to my feet.

My Master frowned but said nothing about my speaking aloud to that goofy shard of mine. He made sure I was able to move without pain, well much pain as I was more bruised than a spoiled apple thanks to all the thudding and raping lately. He went to the door and looked out with worry.

He turned back to me. "Okay, the coast is clear. Follow me in high protocol. I will pretend to be shopping the slave collars. Say nothing, no matter what comes. That my love is a directive."

I took my place behind him then we left the hidden room. I didn't look up from the floor as my Master made much racket speaking with exaggerated excitement upon meeting every Dominant in the place. He showed no shame at his loud announcement to everyone within earshot at our arrival in every section of the Dungeon.

Oh, shit. I forgot to tell you; Master Leo is a girlish kind of schwuler. You know this kind, ja? The feminine talker, walker, oh ja, the queen that is what they call them.

Still nein? Okay think of a bitchy, dramatic male. Oh, never mind. This is not something you would encounter so young I suppose. Trust me Meine Liebe if you ever met one, you'd never forget them.

Then to my horror Mistress Gretta come walking down the hallway. Master Leo turned around and shot me a look of disgust.

He whispered quickly. "Remember my love, not a word no matter what I say." He turned back around then threw his arms wide letting out a loud gasp.

"Oh, meine Gott. Does Aphrodite know her younger sister is on the loose here on Earth? I bet she would be pissed to know you escaped the place she hides you away in. Your beauty is not meant for us mere mortals, Meine heart." Mistress Gretta smiled then looked at the floor upon hearing his over-the-top statement *(and bullshit compliment. Leo is a pure schwuler. He never found a woman attractive in his life)*.

She coyly looked at the floor. "Leo, you tease. A girl can get the diabetes from your words. What are you doing down here in the cells?"

He smiled with pride. "Ah, Gretta meine love, I come here to see the newest crop is all. Okay and to show off my own prize. He matches my shoes, and belt of course, spends much time admiring them on his knees." He threw up his hand and giggled with wickedness at his most untrue sexual innuendo.

Mistress Gretta's eyes went wide when she realized I was following on the leash. "Holy shit Leo, that is the Priceless you're leading around."

Master Leo let out a dramatic yelp, "Nein, get out of town Gretta. You don't mean this old thing, Mad Maxx. Well, I must confess I have heard a rumor he was of the forbidden silver. Since he is now mine, I can confirm that this legend is not correct at all."

The Mistress frowned. "All hype? I should have known Peter was full of shit."

My Master giggled "You're such a negative Nancy, love. I meant the legend is under exaggerated. This boy is a fucking beast the second I get him alone. I can barely keep up with his vigorous demands. I am only one man after all. I am grateful I have three other brothers to aid me. There is just too much on this plate for me to eat all by myself."

I felt my ears burning at my Masters saying those lies about me to this foul FemDom. It took all I had to keep my mouth shut denying his accusations of my wantonness. It was not my trust that he knew what he was doing that stayed my tongue. It was my fear of his quirt. The fact that he had it to keep me from biting my tongue off in the seizure was not a detail I had missed. My ass couldn't take another thudding for at least a week thanks to the now deceased Dungeon Mistress.

Mistress Gretta started to come around Master Leo to get a better look at me. My Master blocked her, still smiling widely at her. She furrowed her brow and backed up with an expression of confusion.

"What is this, Leo? I thought us old friends. I wanted to get a closer look at the Priceless is all." She again tried to get around him and was blocked.

He chuckled with dark humor. "Look with your eyes not your hands, Gretta. This collar belongs to me doll face. I don't want fingerprints other than my own on him. You keep your distance. If you thought Peter stingy you will think me a real hoarder. The boy doesn't want you anyway. He is only interested in the males."

I looked up in pure fury at that bullshit. I nearly opened my mouth to correct this cocksucker when Max pushed me down with much strength. I stared at that motherfucker with hate but kept my mouth shut. after what I saw him do to the killer Christian, I decided it best to let him have his way. I was in no hurry to send the boy into a seizure in front of this bitch Mistress Gretta. I promised myself I would wait and get him and this lying schwuler sonofabitch later.

Mistress Gretta looked irritated. "I hope you are wrong, Leo. I have the rights to his penetration virginity. If he is schwuler then that will be one cold, empty night. I better not have been fooled. Jonas said he is the pansexual not the homosexual."

Master Leo shrugged. "Looks like you bought the pig in a poke. Oh wait, I take that back. You wish you got the poke. With this collar you can only dream since that parts he enjoys are on your chest and not between your legs." He giggled like the schoolgirl while Mistress Gretta grew more angry thinking she had been swindled by Master Jonas.

I glared at Max in shock. He had saved me from interrupting my Master from upsetting the Mistress I hated the most in all the Haus. She was cursing my Master Jonas for lying to her and loudly announcing she would find a way to "trade off" her rights to my virginity. Master Leo had possibly prevented the nightmare of my having to couple with this vicious bitch that had my brother Ryker killed. That fucking soul shard was Gott damned right again, damn him.

Max winced. "Told you. Leo is truly in love with us. He knows you hate Mistress Gretta. I assure you this time, a rose is just a rose. Not everything is an illusion. Sometimes, the answer is simple and the heart pure."

I scoffed then shook my head refusing to believe that this creepy ass motherfucker could love us more than my Annette. He is the man and used us, then got dumped in the plot. He surely was only using us to pay back those that double crossed him. That made a hell of a lot more sense than this crock of bullshit Max was trying to feed me. I began to wish Maximillian would wake up and relieve my watch. Shit, now I was even missing the schwuler boy. What is the world coming to? Help, I am going soft.

Mistress Gretta had worked herself into a frenzy of anger at Master Jonas. "This will not do. I will not be made into a fool. I am going to give Jonas a piece of my mind right this minute. That motherfucker." She flounced with much fury.

Master Leo pulled his hand to his chest dramatically. "Ja, which was dirty of him girl. Look I was headed back upstairs to my apartment. Mad Maxx and I will escort you to his door. That way you have a witness to your complaints."

She smiled, suddenly appearing relieved. "You would do that for me, Leo? You are a real friend, meine love." She kissed his cheek.

My Master wiped off his cheek with vigor, feigning disgust. "Enough. I say I go. No need to torture me, Gretta. Keep those nasty girl's lips off me. You want to make me lose my manhood? I tell you I have a wild boy to please and you go killing my drive."

She chuckled evilly. "Nothing kills your drive, love. Take my arm and let's go slay the Vampire, ja?"

He let out a loud gasp. "Ah, that is such a clever idea, love. Ja, like in the Wizard of Oz. We go to slay the flying monkey." Mistress Gretta took his arm and the two of them skipped – ja they fucking did, the weirdos – like the Dorothy and Scarecrow in that flick. I followed behind them wishing I had stayed home and left the Dungeon Mistresses alone.

Then as we approached the stairwell, I suddenly realized what that clever old schwuler was doing. Mistress Gretta was our witness. I come to understand his plan just as Mistress Gretta dropped his arm and let out a blood curdling scream.

"Oh, Meine Gott. Someone call for the doctor. Heidi and Helga have attacked each other. There is blood. "Oh, help," she wailed loudly while Master Leo started backing up with his hand to his chest appearing stunned to stupid.

"Christ, they are dead, Gretta. I think I am going to be sick. I cannot handle this. Oh, those poor girls. Why? Oh, meine Gott. This is horror. Look away Mad Maxx. This is too gruesome for you, my love," he wailed out just as he turned and embraced me as if trying to cover my sight.

Black collars come running from every direction to the sounds of the two of them shrieking. Then the younger Dungeon Masters come to gawk at the dead old ladies. Soon the place was packed with half the Haus. Everyone wanted to see the "murder/suicide" of the H twins' scene.

Master Leo never let me go, holding me tightly to his chest feigning tears and fear. Mistress Gretta told everyone what happened as she knew it and Master Leo just nodded agreeing with all she said. He would cover his mouth pretending to be gagging and claimed loudly he felt faint from the gruesome sight.

I kept my face buried in his chest to hide my smile at the acting job of my Master. He was damned good, I had to admit. I mean he had fooled me several times with this act of "limp wristed man." I thought of his begging the torture Masters to let him go the day Jonas had him and Peter whipped for interrupting his fondling me. I recalled his fear when Peter claimed he had been masturbating to the blow job I gave to my Master rather than Leo. I suddenly realized he never did any such a thing. I told you I didn't see it. I was busy doing my job. Peter likely was making the whole thing up.

I saw this Leo kill without a shred of revolt. I knew he was ballsy enough to stand up to even the scary Jonas or crabby Claus. He had no problem making a dramatic display of the blow job he did get from me in front of my parents. The man was the panty waist alright, that was made of unbreakable rubber.

I was grateful he was insightful enough to allow me the mercy of not having to express emotion over this bullshit. I am pretty terrible at showing the right one for any situation anyway. Seems I often laugh when I should cry or cry when I should laugh. Worse, sometimes I miss the cues for expression all together. That has gotten me in some trouble at the hospital more than once. I don't know why I fuck up like that, but oh well. Sucks to be me, I guess.

After about thirty minutes of my Master sticking by Mistress Gretta to offer his regret, and disgust over this situation he loudly announced he had enough of "death and sadness."

He pulled me from his chest taking up my leash, then shot a look at Mistress Gretta. "I am leaving this nightmare vision, Gretta my love. My poor heart can take no more. I worked with the honorable Helga during my time in the Dungeons. Seeing her like this tears me in half. I take my leave and will go somewhere of happiness and light to unburden my soul. She is in a better place now. Our loss is the Devil's gain, ja? Bet he is pissed all the work the girls just gave him by doing this stupid thing. Well at least the little silvers have a fighting chance to make it to the floor with these bitches gone." He snorted.

Mistress Gretta sniffed then nodded. "Well, you are right on that one. I won't miss them, that is for damned sure. I never liked their heavy hand. More than a few fine silvers were wasted by their greedy thudders. Do me the favor and have a toast to their eternal damnation for me."

Master Leo scoffed. "Do you see this beautiful creature on my leash, Gretta? The last fucking thing I want to be thinking of is your rat face. Tell Peter and Cora I said suck my ass. I would say it was nice seeing you but the way it ended is not so unexpected. I think in many ways it is even poetic. It is how I feel about you. Our friendship is a dead horror show. As it is, I will not be able to eat breakfast thinking of all three of you." He turned around and pushed through the crowd headed for the stairs pulling the stunned me behind him.

I covered my mouth to hide the smile that spread across it. Master Leo had used Mistress Gretta to cover our tracks, then insulted her once her value was no more. I liked this guy's style a great deal. He was nothing like he appeared. Not even his oldest friend, Mistress Gretta, had learned his complexity and acting skills from her many years of knowing him.

I wondered if maybe I should be careful of his treacherous nature. He had already fooled me completely enough times to make me aware his temperament was the changeable kind. I followed him up the steps reminding myself to keep my eyes behind me to watch for that rapist and for this wily Master's blade in my back.

Once on the first floor we had less trouble getting through the crowds. Most were cramming into the Dungeons. The silvers that passed would drop to a knee as we went by. That made Master Leo chuckle a bit. I as usual ignored them. It seemed very silly to kneel to a nothing like Mad Maxx.

That didn't stop everyone from perceiving me with their own damned illusions. I was long since a victim of overblown fame and false fantasies. That reminded me to keep my alertness to anyone that may be following. I knew the rapist wouldn't grab me while on Master Leo's leash, but maybe the bastard was still watching the show. I looked behind me from time to time making sure to seek out the face of every Dominant in the area. None of them matched the description as I recalled it, not yet anyway.

I was surprised when Master Leo led me to the door. I glared at Olaf and Vilber. They refused to make eye contact while speaking to Master Leo. I examined them closely and determined neither of them matched the man I was looking for. Olaf was under six-foot and heavy set with a light beard. Vilber was six feet, but also too heavy with a moustache. This ruled them both out for good.

To my shock they opened the doors and Master Leo hauled my ass out behind him on the leash. I gasped and began to tremble looking around me wildly for the snipers. Master Leo saw this and smiled with some humor.

"Christian Axel, put on the damned sunglasses and follow me closely. The Guard won't shoot you on my leash. Calm down. Just make sure you don't stray." He stopped till I reached into my pocket and retrieved his shades.

The bright sun was immediately tolerable. I shook with anxiety but did appreciate the mercy of this small favor of the sunglasses. He stood there watching me shake like the newborn calf and then frowned.

"You look gorgeous my love, but this cowering is a painful reminder of the nightmare life you lead. I want you to relax and enjoy yourself with me without you always being ready for a fucking heart attack. I recall my sister had this problem with constant fear and she was well loved. I cannot imagine what it must be like inside that head of yours. You know what? I have an idea. I know something that always made everything better for my Maus. Her name was Sofie, but she was so sweet, Maus was how all of us knew her. He appeared sad for a moment but then smiled once again. Come Christian Axel. Take my hand I am taking you somewhere special." He reached out and I took his hand still scared out of my mind that any minute the bullets would ring out.

We walked across the lawn for a bit then arrived at a parking lot. I saw several parked limousines. There was a black collar dressed in a black suit with a white undershirt waiting by the automobiles. Master Leo walked up and told the man to bring him a "car" that he was going to the Straußenfarm (the Ostrich Farm). The man nodded then took off running.

I stood there unable to breathe from the sight of the cars and his saying he was taking one of them. I had only seen them in magazines or briefly on the television or in films since my abduction six years before. I felt faint and my heartbeat was too fast. Master Leo could feel my palm sweating in his embrace.

He turned to me, appearing suddenly concerned. "Relax my love. This will be fun, I promise. Oh my, silly Leo. I wasn't thinking, my love. I forgot you haven't been off the Grounds in a long time. This must be overwhelming to you. If you are stressed badly, we can go back to the apartment. Just say the word."

I shook my head sputtering out simple words, "Nein, I go with you Master."

Master Leo laughed at my near stupid sounding response. "Ah, good. You will love the Straußenfarm, my love. I haven't been there in ages." A long black limo pulled up to the curb and the black collar attendant grabbed the door.

I followed Master Leo into the back seat near fainting from the thrill of getting the fuck away from the Haus. I was already thinking when we got to town, I could escape this, Master. I would run away and never come back. I assumed I may be shot in my flight but at least I would die free.

I marveled at the sound of the motor and the sway of the ride. I smiled with glee as we passed the fields and I saw sheep and horses in the countryside. Things were so much more beautiful than I recalled. I couldn't keep my eyes off the fast-moving scenery flying past the dark tinted windows.

Master Leo sat there grinning, appearing to enjoy my childish excitement at even the litter on the shoulder of the single laned road. I let out gasps when we passed elder women walking along the path and a man riding a bicycle. I was astounded by all of it. I believed I could die a happy man right that second.

Then the stretched car reached the outskirts of a town. It pulled up into a small grouping of run-down buildings. I was saddened that we would not go through the heart of that village, but I was satisfied I would see it soon enough. When I run the hell off.

The black collar parked the car and I looked for the handle to get out. To my surprise there wasn't one. Master Leo saw me looking wildly for the escape hatch and frowned.

"Ah, my love, you thought you'd take this chance to run, didn't you. I am sad to say that won't happen. It is the law the no silver can leave the Grounds unescorted.

Furthermore, none can be trusted not to try to overtake their Master with such temptation. He got out of his seat and went to the window divider between us and the driver.

I watched Master Leo knock on the glass. The black collar nodded then opened it, handing him a pair of handcuffs. Then he returned to the seat demanding I give him my right arm. I sat there staring at the restraining device thinking this would slow my run but not stop it. I would merely ask to use the washroom and break out of them. I would get away right after that.

I put my wrist out and he cuffed me then to my shock he attached the other one to his left wrist. That made me frown a great deal. Master Leo chuckled at my realization I wasn't getting away that easy.

"I am sorry my love, but you are stuck with your Master Leo. I have your leash and your wrist. Where I go, so do you and vice versa. Now come. I want you to have fun and stop this silly planning to escape. Where would you go? You have no family, no home, no money. You'd be dead the first cold night or killed by vandals. Just let it go and come with me. I treat you today to something you have more than earned a million times no doubt." He motioned to the driver he was ready to be let out of the car.

The driver opened the door and Master Leo pulled me out with him. I was forced to follow him to the middle run-down building. The windows were boarded up and the door was made of steel. The parking lot was full of cars, but all three buildings appeared to be near condemnable. My

Master pounded on this metal eyesore while shooting me a wicked smile.

I began to tremble again wondering if this was a place to torture collars in chains or worse. The door came open and a short heavily bearded black collar looked at my Master with suspiciousness. My Master reached into his wallet and pulled out a card. The squatty man examined it then motioned us to enter.

My Master pulled me along though by this time I was re-thinking my stupidly coming with this guy. The black collar guard closed the door and I saw a short stairway up. It was dark without any real furniture in this entry. I could hear the sounds of beating drums and there were odd colors flashing at the top of those stairs. My Master smiled then pulled me along with him up to that noise.

At the top it opened onto a dance floor. The room was dark with multicolored lights flashing on and off. There was a bar with black collars serving the customers drinks in the very back. All along the walls were tables with chairs. Many other people were there sitting at them, others were on the large dance floor dancing together to a disco beat pumping into the room from huge speakers.

I noticed right away all the dancers were males. I shot a look at the tables to see they also were male. Only a couple of the bartenders and a few of the waitresses that walked the room were females. I looked at my Master completely confused at this point. I thought he said he wanted to make things better for me, that his sister Maus had enjoyed this thing he wanted to show me. She found the schwuler dance club comforting? Huh?

My Master noticed my stunned look and yelled to me, he to as that music was fucking loud, "Christian Axel, welcome to the Straußenfarm. My Maus loved the music and dancing. Ah the woman was graceful as the swan. I have seen you dance. You are beautiful at it. Truly gifted. I would be most honored if you would dance with me. I could live all the rest of my days if you could do this for me this one time."

I stood there staring at him in disbelief unsure what the hell was wrong with this nut. I didn't want to dance with him at a schwuler club, nor anywhere for that matter. If he wanted to dance, he should have just played a damned record at home and ordered it. I would call Maximillian's ass in and be shut of this nonsense. I shook my head.

"Forgive me Master but I don't know These modern dances. I only can waltz or tango. They don't teach this crazy stuff in class. I would beg your mercy to take me home now. I thank you for your offer of this gift, but I regrettably must decline," I shouted back at him watching his bright smile fade to one of discontent.

He nodded, returning to a bitter smile. "Ja, okay, I understand Christian Axel. I wasn't thinking. I should have asked before bringing you all the way out here. I just wanted to give you a surprise." He tried his best to look like he wasn't upset but I could see he was hurt by my turning him down.

I smiled back trying to make him feel better. Though I didn't know why I cared as this was stupid of him.). "Thank you for trying. This is not my scene, Master. It was a kind thought though and I appreciate it." I nearly choked on the words.

Master Leo looked at the bar. "You want to stay for at least one drink before we head back? We have nothing pressing till midnight, ja?"

I looked around at the schwuler black collars inhabiting this club, then raised my eyebrow. "Uhm, I'd rather not, thank you for the mercy, Master." I noticed several of these men were staring at us and whispering to each odder.

I felt a shiver go up my spine. I wondered how long I could survive if these fellows decided to whip my ass for daring to step into their territory.

Master Leo shrugged. "Alright we can leave then. Come on love." He turned to leave but he stopped suddenly before going down the steps.

I was keeping my eyes behind us sure that the whole bar would get out of their chairs to beat us to death. I turned around when my Master abruptly halted like that.

He shot me a look then said, "Shit, there is fucking Peter, Malfred and that cocksucker Grisham. Did they follow us? motherfucker." He turned around wildly then tore off dragging me after him to a table in the corner of the club.

He told me to sit down and keep my head down. I was scared to fucking death. If Peter caught me off the Haus grounds, there was no telling what horror he would pull. I immediately wanted to kill Master Leo again for putting me in this hideous and uncalled for danger. I kept my eyes to the table when the male waiter come to the table and asked for our drink order.

Master Leo ordered two beers and kept his face hidden with a spirits menu. I was hoping in the dark corner that the three Dominants would just leave after not seeing us easily. Then my Master let out a sigh of relief.

I thought he saw them leave but instead he said, "Oh shit Christian Axel I am getting as paranoid as you. They are not here following us. They are trolling for tail."

I looked at him in surprise. "Huh? These are all black collars, Master. That is forbidden."

Master Leo laughed. "Sure, the fuck is on the Haus grounds. This is neutral territory. In the real-world Christian Axel, the rules of the Haus don't apply. You have known nothing but that corrupted place all your life, my love. I understand you think in the way they have brainwashed you to do. The only thing I cannot figure out is why Peter is with Malfred and Grisham. Now the he and Agnette are back together I thought he was satisfied. I mean he did dump Grisham to have her back."

My eyes went wide. "What? Agnette is with Peter, Master? I thought she married this man named Julius?"

Master Leo scoffed. "Who the hell told you that? Agnette wouldn't marry anyone, but Peter and that rat bastard would never ask her. Did you know there is a rumor the two of them have a secret love child together."

I nearly choked at that. "Uhm, okay slow down Master. If she is not married to this man Julius, who the fuck is the guy?"

Master Leo rolled his eyes. "Oh, meine Gott, Christian Axel. This is gossiping 101 with you. Don't you speak to anyone in the Haus? Julius is like Malfred and Grisham, a Voting Council member. Only Julius is a new Dominant to the Haus. He come in like gangbusters from another Haus in Belgium about six months ago. He visited many times before that to secure his seat on the Council. He is nothing but a Nouveau riche family's bratty son if you ask me. He is one rude bastard without any culture. Now that Malfred is one vicious bastard. He is still pissed at the Elders over the tarnishing of his Tamina by some common black collar. I was told he gave her leash out as a favor to Claus, then the black collar got her pregnant. The rumor is he showed up at Claus's door angered with the demons of hell demanding to be repaid the amount of his loss. Until you come along Tamina his silver was the most desired in the Haus for her artistic skill and beauty, but she had gotten older. They say Malfred loved her so much he sold her to another Haus rather than send her to the circuit. Anyway, he apparently has sided with Peter's group while still holding a grudge over that mess about nine months ago now. Ah, that fellow dare, the other big bastard with Peter, he pointed at the man, who is the brutal Grisham. They say he is in love with Peter, poor bastard. The two of them have been on and off again for over three years now. Too bad for Grisham he is nothing more to Peter than Felix was to him."

I gasped. "Felix, you say. Master did Master Peter have an ongoing affair with him too?"

Master Leo chuckled. "Are you kidding? Those two were lovers for over thirty years. Tight as two peas in a pod. Christian Axel you lived in Peter's apartment. Didn't you know of his many lovers or at least of Felix? The man is a fucking animal. He has even managed to get into the panties of Gretta, the lesbian lover of Cora."

I nearly fainted at that. "Oh, meine Gott. I think I need to go back to the Haus Master. I am not feeling well."

Master Leo nodded. "I am sorry Christian Axel. Did I say something wrong? Look I want to take you home, but that Malfred and Grisham are big enough to whoop both our asses and leave the remainder for Peter to sweep the floor with. They work out constantly. I mean look at them. I wish I was that fit. I mean they are in their thirties, and I am old in my fifties, but I didn't look that good at their age for sure."

I turned in my seat at his words "big," "fit," and "work out" constantly.' I noticed both matched the shadow in height and both were clean shaven. I thought of Julius too. He also met the criteria I seemed to recall. They all were associates of Peter, and all on the Voting Council. I realized that only someone that could watch my coming and going would be able to stalk me so completely.

The first rape was by the back stairs, which was a clue I had missed. Only someone that traveled them often would know about that dark corner, and only someone on the fifth floor (the Voting Council floor) would see me every time I left the sixth. Those back steps only lead to the fifth floor and up. My rapist was not only a Dominant male he was most likely one of the Voting Council. This was bad.

I winced then turned back around to face Master Leo, grateful at least now I had a few suspects. Better than nothing. He saw my look of pain and put his hand on mine.

"What is the matter love? Something bothering you? A headache? You seeing an aura? You know light that may indicate a seizure? We really need to get you home for the medication and food. I lost my head with a stupid idea of romance, but I am back to Earth now. I again apologize for my momentary boyish idiocy. I will watch the three of them. The second we can get out of here without their detection we run. Be ready to haul ass. ja?" He chuckled at that.

I nodded, "I am ready Master. I am sorry that I let you down in that dancing business."

He looked sad again, then forced a smile. "That is okay my love. I just thought like Maus, the music would speak to you. My mistake, not yours. No need to apologize."

Master Leo shot another look at Peter and his group of rape suspects. "That fucking Peter. You know I fell for him too, Christian Axel. I was so jealous of you when he was showing you attention. I was so focused on winning his affections I messed up my chances of ever gaining your friendship. I could kill him for that alone. Despite all that, I cannot help it but still love him. The man is so good to everyone, and generous. He has a good heart even if you may not think so. It was nothing personal, I am sure. Peter didn't mean to hurt you, but you were selected randomly." I interrupted him with fury.

"Enough, Leo. That fucker is my father. Agnette is my mother. I am the secret son, Gott damn it. Don't you sit there and tell me it wasn't personal. Mor that the heartless sonofabitch is generous. He and that bitch sold me into this nightmare. He raped me into my fucking collar, Leo. Then used me like a whore. When he thought I was mentally broken he called the Guard to have me uncollared and put to the yard. There is the man you love, fool. Don't believe me? You better think about it. Awful strong resemblance or maybe you chose to ignore that, like everyone else does," I yelled out before I could stop myself.

Master Leo sat there stunned with his mouth on the table while I put my head into my free arm to hide my stupid face. I just yelled at my Master and told him my terrible secret. I was sure my tongue just dug my grave. What the fuck possessed me; I shall never know. I suppose I couldn't take hearing all the praises of my Master for that rat bastard father of mine. Why I cared what Master Leo thought of Peter, once again, beats me.

My Master cleared his throat. "Uhm, I don't know what to say. This is a lot to absorb. You are saying that Peter and Agnette betrayed you too? How did they talk you into their plot? I mean I don't mean to pry but." I interrupted him again without taking my face out of my arm.

"Master, I didn't know he was my father. Felix told me and I walked into the yard that very next day and was shot for it. I was kidnapped from my bed as an eight-year-old kind. My mother told me she did it to keep me from prison. I didn't know any better. I was too ignorant of the ways of the world. Peter came to the dungeons and bought my collar. I never saw the man or heard of his name in my life. I didn't know what a fucking collaring was, nor special services. I was a punk in the dungeons and refused to listen to my lessons. I fell into the traps he laid, all of them. You helped him remember? You must understand I have to break my collar, or this is my life for all time. I never asked for this, agreed to it, nor do I wish for it now. You can go tell the Elders what you know or side with Peter all you like. I am tired. I cannot keep fighting everyone. You say you are my friend. Be my friend then and never repeat what I should never have said to you this moment." I began to weep into my sleeve, feeling the despair come over me like a heavy blanket.

Master Leo got up from his seat abruptly and pulled me from my own. He pulled me into a cuddle then told me to walk with him with my head down. I continued to sob but minded his order. He and I rushed to the steps then raced down them. The black collar opened the steel door for us. We got outside and my Master stood there holding me in his embrace patiently waiting for our driver to pull up.

"Where the fuck are you going, Leo, with that black collar. You bastard, you know the rules, no trysts outside these doors." I heard Peter yelling out from behind us.

I gasped then hid my face in my Master's side as he yelled back, "Fuck off, Peter. What I do is none of your business. Last I checked my level is higher than yours now and my mother is much better looking."

I felt the sun suddenly dim as the three Dominants surrounded us. I wrapped my arms around Master Leo trying to brace for the beating they surely were about to lay on me, and him. I felt my Master tense up.

"Peter, that is a fucking silver not a black. What the hell Leo," I heard a deep voice that I didn't recognize.

Peter scoffed. "I will be damned. Well, Malfred, Grisham, meet the famous Mad Maxx of the forbidden silver class. Leo, you are a fucking idiot. What the hell are you doing taking such a Priceless item off the grounds. What if he had gotten away from you? Damn you are as stupid as you look."

My Master scoffed. "Mad Maxx is secured as the law commands. He has been well behaved. I had no trouble and now I am taking him home. You boys forgive me for not introducing you to my collar, but I think it best he never learns the names of those who will never be of any value for him to know."

I heard the other man growl. "You are the haughty one, aren't you Leo. If you weren't an Elder, I would knock your dick into the dirt, then fuck your silver while you watch me defile him to a whimpering mass of nothing." I trembled at them words and held on tighter still.

Master Leo chuckled. "You are the brute aren't you Grisham? You don't scare me motherfucker. I am an Elder and this is the Priceless of Legend. You lay a hand on either of us, you're all good as exiled. This boy is of particular interest to Claus. I wouldn't piss that man off if I were you. I bid you all a good day. I am taking him home with me. Oh, and Peter, if I ever catch you near him again, I will have you whipped till you pray for death. See you boys later." My Master grabbed me and pushed me into the car that had pulled up blocking the men from even seeing my face as he crawled in behind me.

My Master grabbed my arms and pulled my head into his chest while I wept uncontrollably all the way back to the Haus. I don't know what the hell got into me. I couldn't stop the tears nor the darkness rising within. Relieving that burden of the secret with my own mouth seemed to break the crying jag inside me. I had fought hard to get this far. I blew it all in a single moment, confessing my true identity to the one man on Earth that was known for his shifty nature. I was the fucking king of all idiots.

When we got back to the parking lot. I was done with my sobbing. I felt hollow and lost as my Master had the black collar attendant remove our cuffs. I followed him in silence all the way to the doors. I kept my head bowed low walking behind him on my leash hoping the Guard would shoot me and end my misery.

Nothing seemed to matter anymore. I had lost myself in my fight to get outside. As the Haus doors closed behind me locking me in once more, I realized Master Leo was right. This place was all I knew. I didn't even know how to dance to modern music, couldn't order a meal from a drive through nor had I seen the grass but three times since I was too young to even recall it.

With a heavy heart I admitted to myself I had been pinning for something that was truly unknown to Christian Axel. I had never actually seen the outside world, not at all. I spent two years locked in a barn. Then another six locked in the Haus mansion. I was six years old the last time I knew freedom of any worth. What the hell does a six-year-old know? They can barely count to twenty or color within the lines with a crayon. Everything I knew was from books or beatings on my knees.

I didn't look up as my Master led me up the main stairs to his apartment. He was quiet too, not bothering to respond to the many salutations from the Dominants that crossed our path. When we got safely inside his apartment, I stood against the wall waiting for his next move that surely would result in my being trapped in my metal. Everyone else, but

Jonas, seemed to desire to see that. Why would Master Leo be any different?

My Master sat down on his couch leaning forward his arms on his knees. "Christian Axel, I know you will not trust me, but I swear to you that thing you told me at the Straußenfarm never leaves my lips to another. I swear this on my sister Maus's grave and in blood if that will make you feel better. I suppose you are right. I think deep inside I knew the truth all along but couldn't face it. How can anyone do that to their own blood? If someone did that to my own, but I never could experience that because of what happened to my Maus. I couldn't condemn another to her pain. I cannot believe I loved that motherfucker. The horror you have tolerated, and I am a part of it." He stopped his words abruptly then began to cry.

I looked up with a startle at that. "Master? Please mercy. Your tears I cannot take. If you say you won't tell, I have no choice but to believe you. What does it matter anyway? You are right. What if I do break my collar? Then what? I have no home, no money, no family, no friends, no understanding of things outside of these walls. I am nothing but the metal I wear. I no longer desire to do this anymore for a dream that is all illusion for this dumb boy."

Master Leo ended his crying and stood up. "You stop that. You want to break that collar then I will help you. I was wrong to assume your disease means you should be quiet and just take the abuse of the monsters like Peter, Jonas, or Gretta in this Haus. Well fuck that. I think we forget this afternoon ever happened. I will call down for food, you study, and I will finish my book. At midnight I turn you over to Claus so he can force himself upon you like everyone else always has. Be assured at least four times a month you will be permitted to be yourself without such torments. If ever you need me, I am here. See you do have a friend, and my love, long as I live you will always have a home. Now, what can I order you? I am starving." He smiled while wiping his eyes and walking to the phone.

I shook my head. "I am not hungry Master, thank you though."

He frowned. "Christian Axel you need to eat. You also must take the medications. Don't make me have to get nasty about that. I allow you the freedom to do whatever you like on my clock with those two exceptions. You eat and take the meds, or I have no choice but to take you downstairs to thud. Better your bruised by my quirt then dead from seizure or renal failure from starvation."

I dropped my head and mumbled I didn't care what he ordered. I listened as he called in our meals then hung up and went to his kitchenette singing a pretty tune. When he returned with silverware and sat down on his couch, I realized he was not kidding. This man was being honest with me about everything he told me. How I knew, I could never be sure. I just knew all at once.

"Master that is a nice song you mock. I don't understand the language. What are you saying?" I kept my place against his wall and my eyes to the floor.

He grinned. "Ah. That my love is an American song called "Without You." I sing it in my native tongue. You know what? I have the record if you want to hear the music. I will play it for you and translate the lyrics for you." I nodded with a smile.

I watched him rush to his record player. He dug through his albums a few moments and put on the correct music telling me the artist was called Harry Nilsson. He stopped just before dropping his needle to start the song and looked back at me with an expression of curiosity.

"Christian Axel, I asked you to dance with me at the club, but you said nein. I know the atmosphere was maybe a bit much for such a request. I ask you again. I would be honored if you would dance with me?" He waited for my answer.

I chucked at that. "I do whatever you ask of me Master though I told you I don't know any modern dances."

My Master frowned. "This is a slow dance love. I could teach you the steps easy enough. However, I cannot make you. I ask you to dance but only if you're willing. I refuse to command it. If you say nein, I respect it and think nothing less of you for it. If you hope to break that collar you need to learn to make your own decisions without someone always making them for you. You can never know who you are if no one ever lets you discover yourself." He dropped the needle and the slow song about a broken heart began.

I giggled into my hand when my Master began to dance with a pretend partner across his living room singing to himself. That was one silly sight. I watched him step and glide, caring nothing about my standing there looking at him with humor.

The meal arrived and he put the tray on his coffee table, telling me he had a song I would love also in English. He put one the record of the album called "Days of Futures Past" by a group called the Moody Blues. When he dropped the needle, a sound come across the room that set my Max on fire called "Nights in White Satin."

My Master closed his eyes and began singing this one too. He started swaying across the room like before. In a trance I felt my feet carry me across the room and take his hands in my own. He opened his lids in a startle to see my smiling face in his own. His own smile broke out as his eyes filled with rain. I let him lead me, but he didn't have to say a word to guide me. I already had learned the simple steps of his dance.

As the deep melody echoed on the walls and his handsome singing filled my ears, I suddenly felt connected to this man unlike anyone before. I wanted nothing more than to please him, be close to him. I leaned in and put my lips to his. He shuddered then kissed back. We twirled and deeply kissed, never breaking our dance.

When the song ended and the next began, I noticed I was feeling heated up and panting. My Master ran his hands along my chest, and I found my own interest in pawing back. This was oddly not faked. I found myself desirous of his couple with me. I had lost my fucking mind. Me, Mad Max, was interested in sex with a schwuler? He must have slipped me something I thought, but I couldn't stop the growing desire for his touching.

I didn't even attempt to fight him when he reached for my manhood. He pulled back from our kissing with a sudden look of irritation. My Master broke our heavy petting embrace and stormed from the room. I was left standing there in total confusion unable to get my bearings. What the hell was happening to me?

He returned promptly. I yelped in some fright when he grabbed my arms with much strength re-engaging me in his tongue embrace. I once again didn't attempt to stop him. Honestly, I didn't want to. He and I slowly made our way to his couch. Then my Master gently guided me to my back on the sofa, never pulling away from his wanton kissing of my mouth, ears, and neck.

He unbuttoned my vest then blouse running his hands with much softness across my flesh. Then he moved to my breeches. He undid the buttons and pulled out of our embrace. He pulled them down and reached into his shirt pocket. I whimpered when he removed a hex wrench and rapidly undid the chastity device. He threw the metal monster on the floor.

I shot him a look of terror. "Master, I need to have a virginity for the collar selection. That chastity device protects me from questions. Please I beg your mercy.

Return it to me. I thought you said you didn't want to trap me in my metal." I don't know why I was pleading since I could never find my lust with a man anyway.

Master Leo chuckled. "I swear to you Christian Axel, I will not breach the penetration virginity as dictated by Haus law. You have nothing to worry about. I am a top not a bottom anyway, my love. However, this you are not allowed any kind of release is utter bullshit. There is nothing keeping you from finding your orgasm by manual stimulation except that damned metal contraption. I will never take from you even if you're willing without giving back." He came back at me kissing me down my chest.

I gasped at the strange feelings filling my mind. "I thank you for the offer, but I cannot find my release with the man, Master. If you expect that, then I think we need to stop this right now."

He looked up in a startle. "If you wish me to stop then I will. Just say the word. Anytime you find I got too far say nein and I will let you be immediately. Do you desire I end my adoration?"

I looked into his eyes and shook my head. "I don't wish you to stop, Master."

Master Leo kissed my chest for a moment then said, "Do you say that because you think this is what I want to hear? If so, then be honest with me. I am a big boy. I can handle rejection, my love, but what I cannot live with is being with someone that feels compelled rather than desire."

I closed my eyes. "Then stop this, Master. I don't know what I want or what the hell is happening to me."

He nodded. "I already knew that Christian Axel. Your manhood doesn't lie. Thank you for being honest with me and yourself. For the record, I am flattered you would be willing to try to please me at the detriment of your own interest. I may not be getting my release with you, but I do feel loved. Only someone that loves you would put themselves out to grant comfort to the other." He kissed my forehead and smiled with great affection in his expression,

I looked at him with confusion. "You are not angry at me Master?"

Master Leo got up from leaning over me and began buttoning up my blouse. "Nein my love. This is your Haus while he pointed at my chest. You have the right to refuse any visitors you like. I will always worship you but never desecrate you." I grabbed his wrist as he attempted to try to button another on my blouse.

"Don't Master. Kiss me again and this time say nothing." I stared into his eyes.

"You sure about this, Christian Axel? I thought you said this was not what you wanted." He frowned at me.

I shook my head. "How will I know if I don't try? You told me to explore myself, Master. Until this moment no one ever truly gave me the choice. The closest thing I got to that was accidental. You don't command me, hold me down, chase me, nor tie me up. I feel something for you, but I am unsure what it is. Allow me to find out."

Master Leo smiled then ran his hand along my cheek. "Your wish is my pleasure this time, Christian Axel." He leaned down kissing me with much vigor and lustful interest.

I felt his hand reach down my stomach while he engaged my mouth. To my shock his eager hand found my manhood ready for his mercy, looks like that fucker Maximillian is right. I am the schwuler not him or to be more exact, I find attraction to the heart not the part.

Chapter 44: Leo the Lionhearted Mouse

My surprise at my lustful response to Master Leo's adorations of me was evident to my lover. He smiled upon finding me ready for a mount. His eyes quickly filled with rain, but the storm didn't break through the wail.

He gently kissed my forehead and stopped his pawing with an expression of gratitude and sadness. "Does this signal that you do love me, Christian Axel? I dared not to ever dream such a thing possible."

I gasped in fear at the strong feelings rushing through my chest. I had not expected this shit. "I cannot lie to you. I don't know what is happening to me, Master. I am straight I swear this. Yet, I want nothing more at this moment than to be loved by you and love you back. This doesn't make any sense I confess it. I only hear your heartbeat in my ears and your face is the most beautiful I have ever set my sights on. My desire to possess you is driving me to madness. I think you may be right. I need my medications."

He chuckled at that. "Ja, you need medication for that horrible disease that is devastating your life, my love. I will make sure that happens from this point forward. However, this overwhelming feeling within you is not because of it. I told you Christian Axel, sex is the expression of love for another. You can have one without the other, but when that emotion is pure it takes you it's prisoner. There are no boundaries nor conditions for it. Man, or Frau matters not. You can love both with the same strength and the same joy in the couple too. One should never be so narrow minded as to limit themselves by gender. Natural attraction compels humans to mate and make children. That also can be had without the graciousness of true love. You are a beautiful boy, and I am an unattractive middle-aged man. There was no mercy of the desire for you toward me like I can have for you thanks to that. You belong to me by some silly man-made law in a Haus of lawlessness. My luck in acquiring your collar was your nightmare come to light. Again, another reason for you to want me like anyone wants an STD. Yet here I hold you in my arms and see the passion for me in your eyes and feel it in your flesh. I wonder if maybe I need to take your medication with you." He and I laughed hard at that for several moments.

Then he ran his hand along the side of my face. I trembled and had trouble catching my breath from the electric feeling that filled me from his touching. This made him smile and a single tear broke loose running down his cheek.

I shuddered, feeling I may blow into a million pieces. "Master whatever I did to make you come to tears I apologize. I admit I am not the sexual artist everyone thinks I am. Am I doing something wrong?"

My Master wiped his rain away with a soft chuckle. "My tears are because you are serving me to perfection. I want so badly to believe what I see but because you wear my silver, I cannot trust my senses. You seem to be interested in my couple but how can, I be sure? What if this is only the brainwashing this place has done to you?"

He ran his hand back across my chest sending my eyes rolling into my head. "Nein, Master. I never find lust for my Masters. Not once, nor did I ever expect to. I can fool you, but I cannot fool me. This affection I feel is the real thing. I don't know how you managed this, but I think if you don't make me your own, I may die."

That made my Master howl out in pure glee as he looked deep into my eyes shaking his head in disbelief. "You are in love with me. How can this worthless Leo be so fucking lucky as this?"

I frowned feeling most confused by his behavior of thrill. "I don't know Master, but if this feeling is love then I don't fear saying it. I am in love with you. It is the truth that in intercourse I am well trained but in the art of the heart I am a novice. I didn't even have the mercy of recognizing I was falling for you the second you saved me from Mistress Helga. I feel for you what you grant to me. You are the Dominant of value, worldliness and worth. Why would you be so surprised that this stupid boy could fall for your charms? Forgive my saying so but surely you have been loved by many in your lifetime."

Master Leo sniffed, appearing nearly ready to weep, then he took a deep mournful breath. "I admit in my life I have been blessed with money, fame, and an easy path. Then, I lost the only person that absolutely loved me before you. In my youth, I thought that all there was to life was sex, money, and the jealousy of others. I confess I reveled in it more than you can know because I had all that in abundance. I could afford whatever my greedy heart desired. I tried it all, twice. All the while my beloved older sister begged me to stop being so damned superficial. No one could tell me anything. Not even the only person on Earth that I respected and cared what she thought of me. I would roll my eyes when Maus told me that my existence as the fortunate should have the depth. I would yawn when she preached that our wealth had given us the ability to grant mercy to others. I ignored her time after time. I found the seriousness of reality a bore and the responsibility of helping those not so blessed a burden. Then over a period of eight months when I was nineteen, my Maus come down with the Mother of Madness. Then it was her that begged for the mercy she had granted numerous others. There was no such a return of her kindness granted. For over four years I watched my best friend in the world, nein my very heart, grow weaker. I was helpless to prevent that darkness from consuming her soul, my love. All my partying and frivolity couldn't distract me from the fear I was losing her. Then one day I did. My love, when she died, I found out no amount of money could bribe the reaper to return my Maus. I was a broken man. I found no joy and the pleasures of this world became like ashes in my mouth. I would have given all I had just to hear her kind words to this silly clown one more time. After she was put to rest, I tumbled across the emptiness of my years. I was like the useless leaves of a Spring that has long since fallen to memory. I became the nuisance without any real use, finally ending up here in this lawless Haus. Even the constant drama and fight for survival all around me couldn't break through to my cold dead heart. Then more than two years ago it all changed. Peter called me to enter a plot and I saw you for the first time. A trembling nothing. A scared little boy in a dress. Then I watched as the weak ember of strength became the raging fire. I found myself thinking of your face unable to banish you from my thoughts. Nine months ago, I was told they put you to the yard because Peter tossed your collar. I was devastated for the second time in my life. I went to Peter, Cora, and Gretta in a broken heart fury. I cursed them to a foul death for destroying a boy that did all expected of him even at the detriment of his own comforts. I ended all ties immediately and walked away from them. I no longer wanted anything they could grant me. Not when it had cost me sweet little Maximillian. I grieved for you my love as hard as I had my Maus. Then that night in the Great Hall, the boy of the unbreakable spirit rose from the dead. He came to me and offered to be my own. I knew right away this was my punishment for my part in harming one who never deserved it. I had to hold the thing I had desecrated and look into his sad eyes and wondered why he was so unloved. I accepted this fate. I had earned it. To be granted the reprieve for the crime I committed, the mercy of it. At last, I understood what my Maus had tried to tell me so long ago. I realized this is my last chance to do the right thing for once in my sorry life. I would use my blessing to protect the cursed. I can promise that I will never fail you my love, nor turn my back when you call. I am yours till they feed me to the buzzards. That I vow. I know I don't deserve your love but that doesn't mean I turn it away when such a gift has indeed been granted. I have to say I find it ironic that after all the wandering I find my living heart in the exact situation that I left it. I can see the sunrise in his sky-blue eyes. The schizophrenic named

Christian Axel that has never known safety, freedom nor kindness is the love I had been seeking all along. I grant all that I have or ever will be to you without restraint or regret. I know my Maus is watching me now with pride. At last, her Leo found his depth. If she can hear me let me say to her, she was right. Sharing your soul with another is the greatest pleasure I have ever known."

He leaned down and kissed me showering with the tears of his emotional overload. I reached out and grabbed the sides of his head. I clutched his hair fearing letting him go. A sudden trembling ache filling me. I couldn't stand the idea of ever leaving his embrace. Every molecule and cell of the flesh cried out his name.

My Master returned to his eagerness of undressing me. He gasped, nearly falling over in a spasm of thrill as I ran my kissing and licked his neck and ears. He groaned in frustration when he was forced to pull me off him so he could undress himself in preparation for our lovemaking.

I watched him never taking his eyes off me as he tore open his own shirt and ripped off his jeans. My Masters expression seemed near mad with lust. Once he was fully disrobed, like the Lion with a loud growl, he pounced upon me making me laugh out loud at this silly act.

He rubbed himself on me with much carefulness not to crush me under his weight. I wrapped my hands around him scratching with wantonness practically begging for his penetration. Ja, do I need medication or do I? My Master moaned with much passion when I nudged him to roll taking my place on his own back.

Den with all the skill of my couple years of enforced sexual services I began our coupling. I kissed down his chest, stomach and finally began my oral adoration of his manhood. My Master panted and called out just as loudly as he had spent the day in the dungeon room with my parents. I fought back the urge to laugh realizing that was not the act I thought it was. Master Leo was a screamer.

Master Leo then grabbed my shoulders lifting me back to his mouth for deeper kissing than I had ever encountered. The passion was building causing my own interest to rise once more. My cock was grinding into his belly. He pulled away from my lips then looked at me with some shock.

"That is a fucking monster you have there, Christian Axel. Thank Gott I am not a bottom, or maybe too fucking bad I am not. Holy shit, that is an awful lot of Christian Axel you are packing there, my love." He looked down at my manhood with that same expression I saw the day in the laundry room from the trainees when Ryker had me put on the dress.

I looked down with a cat smile. "I keep hearing that, Master. I apologize but nothing I can do about it."

Master Leo broke out in loud laughter then grabbed me spinning me to my back once more. "Nein, there is not a thing you can do, but Leo can." He winked, then to my shock he dropped down and demonstrated his own oral skills to me.

I knew I should stop this breach of my penetration virginity – ja, that wasn't really there thanks to my sweet Annette – but I no longer cared. Master Leo had enough of me to trap me in my metal anytime he liked. I thought to myself if he was going to betray me, may as well get something for it, ja? It didn't take him long to send me to a roaring climax. He didn't shirk from his duty to attend me to my full satisfaction, nor did he spit what he worked so hard to obtain.

I was panting out, feeling I may faint from the incredibly powerful release when my Master returned to kissing me. He chucked in my ear when I had to keep pulling away to breath thanks to my relief of my pent-up urges.

"Damn, Christian Axel. When was the last time Jonas took off that fucking chastity device? You act like you haven't cum in twenty years," he whispered while kissing my neck.

I shuddered. "I have been in the damned thing for quite a while but that was my first real blow job Master. I apologize. I didn't, uhm, expect it to be so awesome."

My Master pulled back appearing stunned. "Seriously? You never had a, oh Fichen mich. Those dirty bastards never give you such a mercy. Shame on them all. For all you have done they owed way more than that. Well, never mind. Your Leo is here now. No more of that one-sided bullshit. I know Jonas does this to keep you from spilling your seed. That dumb bastard thinks you will lose your powers that way and be unable to transfer them to your female Priceless mate. Pure fantasy I tell you. Besides, you will need to marry a Shetland pony with a cock like that. I am kidding. I tell you, my love, your Frau will be a woman that never stops smiling or will be one fast runner." He giggled at those last few things he said.

I ignored his teasing but was startled at this latest indignity information. "Huh? Master, you know Jonas does this for that reason. He keeps me suffering because he believes that having an orgasm lowers my Priceless powers for truth?"

He nodded while running his hands through the top of my hair. "Ja, sadly that is why. The man is the nut. He really believes in that legend of the forbidden silver shit. I am wondering if he is not mentally ill himself. There are no fucking powers in being a schizophrenic. If anything, it steals your abilities, it does not enhance them. His false beliefs will injure you my love, and maybe the poor little female silver you level with your own hands. Never mind the madness of that idiot Jonas. You want to cock block me, Christian Axel? Now, I would like to continue this amazing couple with you, but you need to eat and take your meds. At the risk of losing my own release I say we end this fooling around for a bit, lets attend these other things, then see what happens, ja? I doubt your healed enough for my penetration anyway. I am in no hurry my love. Half the fun is in the wait." He sat up and stretched.

I sat up in shock then looked at his clock. "Master, it is ten thirty. I need to go to Master Claus and Bladrick in only an hour in a half. If you don't finish this intercourse now, it will be seven days before I am back this way." To my surprise I was upset over the length of time I would have to wait to hold my Master once more.

He smiled then leaned over and kissed my forehead. "Ja, I know, my love. You're in need of a different kind of care from me, something besides my lustful thrusting. I grant you the mercy of not enduring the pain of penetration from your lover. Let Claus be the brute because I refuse to be the non-empathetic bastard. You have a chemical burn for fucks sake, and three days not even close to the time needed to heal. When the great day comes, we consummate our love. I want it to be without pain and only with pleasure. For that reason alone, I give you the release you more than deserved. I can hold off getting my own. I told you I am not a beast without control of my base urges. I won't lie and say it doesn't kill me to not be spending the night in your Haus while wrapped in your arms, but I will live."

"Thank you for your kind mercy, Master. I will never forget what you have done for me this day." I smiled then leaned in kissing him with much gentleness.

My Master groaned as he pulled out of my latest tempting of his interests. "Come have dinner with me, then I watch you take those pills. Hurry up and redress and let's do this. I warn you my love. If you don't mind me in these only two things I ask, I will have to punish you for denying my simple to follow commands. In fact, I make eating something every day and taking your medications a directive. You know what? I keep these pills. You come here every day and I will give them to you. That is the only way I can be sure your illness doesn't make you dishonest with your Leo."

I frowned at that but quickly put my clothing back on after waiting patiently for my Master to re-apply that metal monster. He growled and cursed my Master Jonas under his breath for demanding such a thing be worn in the first place. He then collected his own shirt and jeans off the floor. I dropped to a kneel then aided him with his task of redressing. He made many compliments over his perception of the grace of my years of training. My Master kissed my forehead and hands while I attended him with much adoration.

I took my place at his feet to follow his order to eat the meal. He shook his head then demanded I join him on his couch. I sat next to him choking down some of the meal. It tasted like metal and dust. I coughed and felt sick to my stomach. Master Leo watched me with an expression of concern demanding I finish the whole plate.

I did as he told me, feeling frightened that Master Jonas had tainted the meal. When I got all that nasty food swallowed, I begged to be excused. I rushed to the washroom and forced my vomiting of all that trash up. I prayed I managed to get all the drugs out of my system but assumed some may have leaked into my tissues.

I was stalling in his bathroom trying to avoid the pills when I saw Maximillian and Mad Maxx show up. They had come to take the wheel for their shift. I was not really happy to see either of them. I wasn't ready to leave Master Leo. So there, I said it, sue me.

Maximillian stood there his expression full of shock. Mad Maxx's eyes were wide as well. I glared at them for a moment but then dropped my gaze to the floor with some embarrassment at what I knew they had seen.

Maximillian scoffed loudly. "Mad Maxx, do you believe this shit? Our bad ass sadistic brother is a fucking schwuler. A real one. I knew it. What happened to all that wanting to fuck Annette? Guess if you cannot get the pussy you think to just become one? For fucking Leon, brother. What the fuck is wrong with you?"

Mad Maxx nodded. "Mad Max, I have to agree with Maxmillian. You kicked out Christian so you could save your schwuler lover? If I hadn't seen it, I would never believe it."

I felt my ears burning with anger. "Fuck you both. Leo is wonderful. You're speaking without any Gott damned understanding of this." Truth is I didn't really know what the hell was happening to me either.

Maximillian laughed loudly. "Ah and this guy calls me the whore. Well, for the price of a trip in a limo, he gives the blow job. Shit, I guess that sucking cock is okay if you get paid. I suppose I just need to raise my rates, then I can call sex with the man I am forced to endure love too."

Mad Maxx laughed. "You know what, brother? I think since Mad Max loves to fuck men we leave this homosexual to deal with the nasty pawing of Claus. He loves the blow job so Bladrick's bloody urine will be his pleasure. ja?"

I winced while they high fived on that shit. "I do not love to give blow jobs nor be fucked by a man."

Maximillian snorted then laughed "Oh? Then I was watching someone else licking Leo's dick like a lollipop with eagerness? Brother Mad Maxx, help me, I am told I am hallucinating."

Christian showed up growling with his eyes red as blood. "Nein, you don't hallucinate brother. This schwuler fuck and that sissy playmate of his would have squealed like the Frau if Leo had fucked him. Hello there cocksuckers, remember me?"

Maximillian crossed his arms. "I think we need to call Der Hund. Mad Max has gone mad. I vote we shatter this motherfucker before he gets any worse. Next, he will be seeking out our mother and father to fuck with like the slut he is."

I shot a look of hate over that uncalled insult right there. "Enough. I will not answer the likes of any of you. I dare you to try to shatter me, your heartless cocksuckers. I can love Leo if I wish. There is no law against it."

The shards began to suddenly grow larger. I backed against the wall feeling fear rising at this horror show.

Maximillian smiled at his increased power. "Ah, look at that, brothers. Der Hund just gave us strength, but he seems to have overlooked our schwuler the Mad Max. I think we have our answer. Get ready to be sent to oblivion brother. We cannot have this weak link in our chain. He compromises the mission with his illusions of love. Get him."

I let a wail while the three shards came running at me for their attack. The boy fell backward cowering in terror. I knew I couldn't beat all three of them even if they hadn't gained that extra girth.

To my shock, they bounced backward and were flung into the walls just as they tried to enter the flesh to destroy me. I didn't move from my spot on the floor, covering my head trembling. Mad Maxx got up and ran at me again. The result was the same. The Shards couldn't penetrate the boy's wall to finish me off.

Maximillian growled with fury. Mad Maxx glared with hate. Christian got up from his latest fall and shrieked till I covered my ears from the pain of it.

He stood before me finally ending his inhuman screeching. "Come out of there you cowards. You cannot hide out in the boy forever Mad Max. The second you step out; we are going to rip you apart."

I began to loudly vail out, weeping uncontrollably. "Nein, nein, leave me alone. You don't understand. Help me. This is a nightmare, nightmare, nightmare, heightmare, tightmare, rightmare, frightmare, mightmare."

The door swung open, and Master Leo rushed in to find me rolled into a ball rocking and chanting this way. My screaming had alerted him to another fit coming on. He rushed to my side checking to see if I would respond to him. When he found me stuck in a loop, he ran the cold water in his tub.

Without any gentleness he hauled my agitated half crazed flesh to the faucet then forced my head under it. The shock felt like electricity running down my spine. I lit up with the fires of hell everywhere. I howled in terror then tried to claw at my attacker with all my strength.

Despite my aggressive scratching Master Leo wouldn't relent. "Be still, Mad Maxx or I will take you to the chains. Do you hear me? I will beat you till you cannot stand if you don't stop this screaming this second. This is a fucking directive."

I whimpered out but understood the threat. I immediately halted my insolent behavior. My Master didn't let me out of the facet for another few moments. When at last he believed I was willing to comply, he pulled me out of the water. He grabbed a towel and roughly dried my head. I sat there trembling in a trance feeling agony unlike any I had ever known. My brothers railed and gnashed their teeth all around the room.

The sounds of everyone in the Haus came through the floor. I could hear what everyone was thinking all at the same time. It was deafening. I covered my ears and dropped my head into my lap pulling up my knees in silent agony.

Master Leo watched this with fear in his expression. "Mad Maxx, listen to you Master. This is not real. You are hallucinating. Come back to me this minute. Don't you give into those illusions. Follow the sound of my voice. This hearing others than your Master is not reality. You need your medication."

I began to cry loudly, full of terror at all the noise. The light hurt my eyes. I had to get away from this torture. I looked up with much startle watching this man kneeling next to me move his mouth. I couldn't understand the words he spoke. He seemed angered. I couldn't find my location in the room. I was everywhere but not anywhere.

Maximillian whispered, "They are trying to trap us in the metal Mad Max. Run. Run."

Christian growled in fury. "Kill him, damn you. Kill all of them. Burn this rats' nest to the ground, motherfucker."

Mad Maxx groaned out in pain. "We are doomed to the chains brother. You must hang yourself to end this torment."

I couldn't get them to decide. I shot a look of fear at the strange man in the room then got up and ran for the door. I had to get out. I rushed down the hallway with the floor shaking under my feet. I made it to the living room when I fell face first to the ground. A weight was on my back. I groaned in agony thinking the ceiling must have fallen on me while I attempted my escape.

I could hear the drumbeats of the schwuler club. I felt myself rolling without trying. Shadows hoovered over me. Then my mouth flew open, and it began to rain. I tried to close it before I drowned but it was stuck.

Suddenly it snapped shut with my mouth full of the storm. I shook my head trying to spit this out before I smothered. My nose stopped taking in air with my nostrils slamming shut. I swallowed the foul liquid to attempt to breathe. My mouth came back open. I wailed in fear that I was pinned to the earth unable to move. This had to be hell. I must have died.

My panic was rising. I fought the huge boulders that held me hostage from trying to get away. A soothing sound began to float through the air.

I stopped my struggle to listen as words I could not understand called out to me, *"Nights in white satin. Never reaching the end. Letters I've written. Never meaning to send. Beauty I'd always missed. With these eyes before. Just what the truth is. I can't say any more. Cause I love you. Yes, I love you. Oh, how I love you."*

Another voice joined it. "Sleep Mad Maxx, my love. You are safe. Your Masters are here to watch out for the bogie man. You are tired. Let the medication make the pain go away." Butterflies began to light on my forehead.

I whimpered in fear that I was falling asleep. I couldn't dare to lose my alertness with this attack going on. I would surely be killed. If I wasn't already dead, which was pretty likely I thought. The pain was so terrible I hoped that if I was still breathing that it would stop. I didn't think I could handle this kind of agony any longer.

I awoke to the sounds of an alarm clock. I was in a bed. I sat up with a startle. Master Leo jerked up and grabbed me around my chest hugging me tightly. I flinched till I saw who was grabbing me.

I smiled then lend my head back into his chest. "Oh, I didn't realize I was in your bed Master. Thank you for the privilege." I chuckled at that.

Master Leo didn't relent his tight hold. "Christian Axel, do you know who I am?"

I sighed. "My lover?"

He let out his breath slowly sounding relieved then spun me around. "Look at me, my love. You had a bad episode last night. Do you recall what happened?"

Ma eyes went wide as I realized something was wrong. "Wait. Where are Master Claus and Master Bladrick? Did I dream the whole thing? The Straußenfarm, the dancing? Are you not my lover Master?" I began to feel somewhat relieved thinking the whole thing had been a nightmare. I am not a schwuler. How silly.

Master Leo smiled. "Nein. We did go to the Straußenfarm and came home to spend time as lovers. You went to the bathroom, and I found you in a psychotic state, my love. The Haus doctor had to sedate you and your medication forced. Claus and Bladrick will be checking on your status. I told them to leave you be, but the cads mind me like they do their farts. Without interest and complete rudeness." He wiped my chin of the leaking drool with the sleeve of his shirt.

I gasped. "I had a fit you say Master. Another seizure?" I trembled realizing all the frightening things I recalled had happened. It was no nightmare, well, ja, it is but a waking one."

He nodded. "Ja, a seizure too. My love, you must take your medication. This missing the doses caused this rebounding of your onset. Do you recall my only two directives?"

I nodded. "I must eat when in your leash and return daily for the medication, Master."

Master Leo smiled with happiness. "Ah, that is correct, my love. I shall give you a reward for recalling my orders. You have earned a palace, but a kiss will have to do ja?"

I closed my eyes as he leaned in, giving me a deep, passioned filled kiss. I felt my flesh heat up with interest as it had the night before. There was no longer any doubt. I was head over heels for Master Leo. Hopelessly in love with the man to be exact.

I returned his kissing and ran my hands across him pushing him back to his pillow. He offered no resistance.

He wore only a button up shirt and jeans. I too was completely clothed. His being honest and not taking advantage of me while I was unconscious from sedation was not a detail I didn't appreciate. It also proved him a man of his word. That only made me love him more.

I had started another heavy adoration of my Master and had only just removed his shirt when a pounding began on his apartment door. He groaned out with frustration as I kissed his chest faster hoping he would just ignore the visitor.

"My love, we have to stop. That is Claus and Bladrick. Fuck, tell you what. I go kill them both and return to you. Then we stay in this bed for all the time we have left until the Guard comes for us?" He smiled then chuckled with me over the silliness of such an idea.

The knocking came louder and with more fury. "Coming, coming, you old goats," he yelled out the first words but whispered the insult only loud enough for my ears.

I looked at the floor. "Can you not send them away, Master? Tell them I am not recovered from my seizure? I can pretend to be out cold." I wanted nothing more than to stay with my Master Leo forevermore.

He smiled with bitterness at that. "Oh, what I wouldn't give to do exactly that, my love. However, you must break that collar. If you push it these old buggers will lose interest, I fear. It is not right, but it is the truth. I hate them for what they do, and I hate myself for not being able to end it. If I tried to run with you, they would stop at nothing to hunt us down like dogs. You earn being a Dominant, then we can be together all we like without fear of them. You are not alone anymore. Leo is with you. You do this for you, do it for us. I will always be waiting to grant any comfort in your brutal world that I can." He stroked my cheek.

I grabbed his hand and kissed it. "I thank you for loving me, Master. I want you to know I love you back. I have explored and found myself in you."

He closed his eyes nodding in rapture. "Call me Leo, Christian Axel. I walk beside you not ahead. We are lovers, friends, partners, and above all one heart. In front of the others, you can use that nasty word, but in my embrace always tell the truth. I am Leo, not you Master." He then kissed me once more with much passion.

The knocking began again in earnest. He pulled away and jumped off his bed with much irritation. "Motherfuckers cannot wait to take my beloved hostage, then rape him. A fucking sick world, just revolting." He muttered to himself as he rushed to his door.

I leaned back into the pillows wondering how this weird situation had happened. Then I gasped as I saw Der Hund walk into the room. I shivered with immediate terror. I was sure he come to shatter me for fucking up the Christian bond, then falling in love with Master Leo like the Gott damned fool I am.

Der Hund smiled; I think. His mouth is a real mess, you know. "Calm down, my sadist. I don't come to harm you. I come to straighten out the problems my judgement has caused all of you."

I frowned at him. "I don't understand."

He chuckled then called Maximillian, Mad Maxx, and Christian to him. My brother shards came into the room, then knelt at his feet. I got out of the bed. Max and I did the same in the flesh. He took a long, deep breath.

"Mad Max does my bidding just as all of you do. To question his behavior is to dispute my authority. I approve of this love affair with Master Leo. In our brutal lives we have never felt tenderness nor care from another. This Master offers it with purity and honesty in his heart. The second that Master Claus and Master Bladrick take the boy, Maximillian you will exchange the wheel with Mad Max and my soul Max. Mad Maxx you will join your brother and offer support if needed. The moment those schwulers have filled their lusts, Mad Max and Max will return to the boy. I am ready to temper the violence of the flesh. Christian you are banned from riding the boy until further notice."

Christian growled out in insolence. "This is not fair. There is a rapist to kill, and we haven't even started on that fucking Voting Council. I also intend to kill Gustov. I liked Abelard and that monster fucked him over so bad he committed suicide by Guard. Nein, I will not mind this order, Der Hund. Fuck you." He got up and stormed from the room.

I nearly choked on my tongue in terror. Der Hund shot a look at the rest of us. Then tore off after him with fury in his eyes. I looked up at Mad Maxx and Maximillian in fear.

"Der Hund will shatter Christian for that," I said with a tremor starting in my hands that startled me a bit.

Maximillian glanced up at me. "I apologize for calling you a schwuler slut though you are one."

I snorted. "Takes one to know one, whore. Enjoy that threesome you got coming. Good thing you don't have another hole, or the Elders would invite Master Jonas to plug in. You will love it prissy boy."

Mad Maxx scoffed then shot me a look of surprise. "Wow, you say that to Maximillian when you just practically raped Master Leo with your mouth. Seems to me you're the one that loves it."

I laughed with much wickedness. "Your just jealous brothers. I know it pisses you off that Master Leo is all mine. You both get stuck with the trash while I take the treasure for myself. I am the one that found him, so I get to keep him. You two get sliced and sucked or knock the dust off. I get to dance, go outside, ride in limos, and enjoy the orgasms with expert blow jobs. I must admit this is quite the burden, but I am sure I will find the strength to endure all the loving, kissing, freedom, and tender words."

Maximillian looked up with shock in his eyes. "Ficken mich. He is right, brother. Gott damn it. The sonofabitch may be the schwuler, but that motherfucker gets the cuddles while we get the blood curdles. This is bullshit. I want to trade with you Mad Max. Give me Master Leo. You can have Master Jonas, or any of them for this one."

I shook my head. "Forget it. You heard Der Hund. This Master is to be served by the Mad Max and Max. Not the flotsam of the schwuler loser, brothers."

Mad Maxx growled out with fury. "Stop fucking calling us schwulers. You're the one in love with a man. You desire the couple with that Master. You are the cocksucking schwuler, not us."

I really giggled at that. "Damn right I am going to enjoy the intercourse with Master Leo. You call me the schwuler, but I also coupled with Annette and would do it repeatedly till my dick falls off. I am not a schwuler. I am the Pansexual, fools. You boys on the other hand have never fucked anyone, frau nor man. You get fucked by the men and you hate doing it. I have been well loved by two. One of each. I will allow you to call me lucky, or maybe greedy, but a schwuler? Nope. I still wish to fuck all the pretty girls and I only desire the attentions of this one man, Leo. Anyone else, yuck. Not interested."

Maximillian dropped his head realizing my words rang true. "Mad Max I right. I always get all the dirty work. This fucking blows."

I nodded. "Ja, you will blow Maximillian. You too Mad Maxx, and even me the sadist Mad Max. Face it boys. We suck cock like pros and then thank the Masters for the Gott damned mercy of it. It is our job as a fucking pleasure submissive. Get over it. You want the choice to say nein, then we need to earn it. Back to work brothers. This collar isn't going to break itself." I chuckled at that, then headed to the washroom to attend the necessary hygiene for the flesh's special services duties.

I heard my Master knocking as I finished my prep. "Mad Maxx, love, Claus and Bladrick won't listen to reason. I will have to let them leave with you, but they did agree to let me attend you if there is any trouble. We didn't alert Jonas to your little fit. I ask you kindly to not inform him," he called out from behind the wooden door.

I opened the entry with a smile greeting my agitated Master. "Ja, I agree he need not hear of it, Leo. I expected that my Masters would not relent long as I beg, scrape, and bend over. I have been in my collar long enough to know the resolve of a Dominant to possess what they want when they want."

He nodded looking at the floor with sadness, "I suppose you do indeed my love. Well, I suppose this is goodbye for now. I see you tonight at the Great Hall though. Claus and the brothers want to go as a group. I was happy to go with them if it means getting to gaze upon my beautiful Christian Axel."

I wrapped my arms around his waist pulling him close. "Then I look forward to tonight. I beg you not to be tormented by my situation. I am accustomed to the pain of it. It barely bothers me anymore (boy there was a fucking lie). No matter what they do to me you're the only one that will be on my mind and in my heart. Ich liebe dich." I leaned into engage him in kissing when suddenly Max and I were tossed to the floor with much violence.

I watched Maximillian and Mad Maxx as they got the wanton kissing from my lover. They had hijacked the boy several minutes too early on purpose. I was pissed beyond imagination that bastard stealing the affections from Leo that is exclusively mine.

I got off the floor storming back for the flesh. I was gonna kick Maximillian's ass for this foul trick. Max drug his feet and held me back with great strength, preventing me from re-entering the boy for my revenge.

I turned to that cocksucker. "Let me go, Max. Maximillian is kissing my Leo. That is not his role."

Max shook his head. "Let this be Mad Max. What does it hurt to let the hardest working shard have a thrill for a moment. You need to learn to share the love as much as you wish to loan out the burdens. This gives your brothers a reason to fight on. Right now, we need all the aid we can get. Christian is on the loose, there is a rapist stalking us, Peter is threatening to retaliate, and the Madness looming. I will not allow you to fight your brothers. More drama we don't fucking need."

Maximillian and Leo broke from their lustful embrace. That fucker schwuler shard looked over and winked at me. I swear to Gott I wanted to beat him to death, but Max held me to my spot. Thankfully, Master Claus and Master Bladrick grew impatient. They began to yell from the living room to "hurry up."

I shot a smile at Maximillian when the Elders called for him. "London bridge is falling down, falling down, falling down. London bridge is falling down my fair Maximillian. Better hope the bridge doesn't fall down, or it will get a thudding. Ah, no worries brother. It could be worse, ja? You should be grateful there are only the two places for their pleasures, you lucky boy. If you had been the girl, then TP wouldn't be short for toilet paper in your world now, would it? Don't forget to count motherfucker. See you soon sister, uh I mean brother, sorry fooled me a second there." I laughed wildly when he secretly shot me the bird while picking up our travel bag to ready himself to leave.

Quick note: In case you forgot/didn't know, Mad Max is referencing the center spot in his threesomes of Maximillian on hands and knees with one partner behind and one getting a blow job. That is called the London Bridge position. TP is short for triple penetration if you hadn't already guessed it. Mad Max is back to being a rude fucker as usual. So much for love changing a personality, huh?

I ignored Mad Max's taunting. I was having a lot of trouble believing he had landed us in a love affair with the schwuler Elder Leo. Though I had to admit his affectionate kissing and romantic words did make me feel oddly good. I felt compelled to try to stay in his apartment as long as I could. I wanted to experience this amazing sensation within me a bit longer than Mad Max had. It simply wasn't right that I always got used as the pincushion without any pleasure from it. This Master Leo seemed willing to at least return the Gott damned favor for what he intended to take.

My Master took up my leash and politely asked me to follow him. He reminded me that in front of the other Masters it was back to Master Leo and best we hide our true affection for one another. He told me Master Jonas and the Elders would not understand. I agreed. I had already witnessed what jealousy could do. Ask Helga and Heidi, yikes.

I nodded. "I will do as is asked of me Master. You asked that I return for the pills. Is there a time? I think it would be best if you shared that directive with Master Claus and Master Bladrick. Otherwise, I may be stuck unable to reach you for this mercy you offer." I didn't want the pills, but that nightmare the night before wasn't something I wished to suffer any longer either.

Master Leo sighed as he opened his bedroom door leading me out into the hallway. "You need them every four hours, my love. I will tell them but all I can do is hope those idiots will hear me."

We entered the living room to find the Eldest of the Elders sitting in Master Leo's chairs. They looked up at me with wide smiles on their faces.

Master Claus spoke, "Ah there he is. Our beautiful Priceless. Isn't he the most beautiful thing you ever set your old eyes on Bladrick? A true vision of perfection."

Master Bladrick nodded. "I agreed with you, brother. I admit I have been most eager for this day of holding him for our own to come at long last. I worried I would be pushing daisies before it came, thanks to that fucker Drexel and this silly stress reactions. I for one am grateful Barnim and Drexel were clumsy motherfuckers. I don't miss them."

Master Leo cleared his throat nervously. "Uhm, ja, so brothers Mad Maxx is in need of medications for his illness. The pills must be taken every four hours or the shit you saw last night will rapidly come on again."

Master Claus shot a look at me, and I looked to the floor. "That is fine. Give us the stuff and Bladrick and I will set an alarm. Not a problem. When is the next time for a dose?"

Master Leo shook his head. "Right now, before he leaves. brother I would ask you both to forgive me for asking a favor so soon after your kindness at raising me in this Haus, but I need one."

Master Bladrick scoffed then crossed his arms. "Well, asking a favor upon a favor is not so uncommon around here. What the hell do you want, Leo? You already have more than most anyone could dare to dream."

Master Leo looked at the ceiling. "You got me their brother. However, this favor is not for me really. It is for Mad Maxx. I would beg you to grant the mercy of making sure he gets his required medications on time. I need you to allow him to come to me every four hours during the daylight beginning right after the wake-up service and just before turn down at bedtime."

Master Claus growled. "What is this you ask? Do you insult us by wanting this task thinking we are too fucking demented to give the boy his pills on time." He sounded angered.

Master Leo shook his head and put up his hands to his chest dramatically. "Oh goodness, nein. Brothers, it is no secret my sister Sofie suffered this disorder as does our collar. The schizophrenic tends to shirk their medications and can be quite clever at deceptions in avoiding taking it. I have experience with their tricks, and avoidance of this necessary task. I wish only to alleviate you from having to thud the boy when he pulls his antics driven by the symptoms of his disease. I mean no insult to my most honorable and capable brothers."

Master Bladrick stood up with some slowness. "Let damned Leo give the boy his pills Claus. I for one don't give a shit how he gets them or who gives them. As long as our collar is being treated right and being cared for, does it matter how it is done?"

Master Claus nodded. "I agree, Bladrick. Fair enough Leo. I ask you to give the curtesy call to our apartment when the medication is due. We release the collar to come for his medications, then return to his service to us right after this task is done. That is my ruling. This will affect Jonas as well. I will tell him tonight at dinner. He will be most angered by his loss of control on this, but that only gives me more thrill in making this a new rule for our shared collar. In fact, while I am on this subject, I think Jonas gets too much time with our shared prize. You know this illness he has. I was impressed by your quick and skilled treatment of his psychotic fit last night, brother. I hereby rule that of the six days left after our two in every week you and Jonas shall share three of them equally."

I shot a look of pleased surprise at Master Leo. He turned to me with an expression of pure bliss on his face. I swear to you Meine Liebe, I felt my heart speed up with excitement. I suddenly understood why Mad Max wanted this man so badly to even overlook the burden of being penetrated by him. I wanted nothing more than to run into his arms and be his for all time. I couldn't believe I was feeling so wanton for a man, a Gott damned Master and fucking Leo of all the people on Earth.

I stole a glance at Mad Maxx and saw he too was falling under the spell weaved by this Elder. He didn't have to say a word for me to know that he was also lost in Leo's eyes. I trembled with weakness in my knees when Master Leo reached over and quickly squeezed my hand.

He nodded. "I go to get your pills, then you go with your Masters, Mad Max. Mind them well or I will hear of the reasons why my love. You remember what I told you. Eat your meals and no stress. I expect you back here in four hours. No tricks or down below we go." He did his best to sound tough, but I could see the rain of happiness behind his eyes at the gift Master Claus had just granted.

He come back with a glass of water and those mind controlling drugs. I had a moment of paranoia thinking maybe that is why me and the brother shards were falling in love with him. I took them and started to try to hide them in my mouth to avoid swallowing when Mad Maxx tapped my shoulder.

"Take them, brother. If this is what is causing this amazing feeling for Leo, then why not? All we have ever known is pain. This man makes us feel joyful and safe. If all it ever took was magic pills, then I say take the whole fucking bottle. I don't want this love we have for him to go away, and neither do you." He smiled with a calmness I had never seen in him before.

I nodded then drank the water and took those mind-altering things with honesty. Mad Maxx was right. If they were behind our interest in Master Leo, so what? The way he made us feel was the greatest thing in the world. I would be willing to steal, fight, kill, even die, to keep the glow of it in our heart for the rest of our life. Everything could be perfect, if we could just find a way to knock Mad Max out of the rotation so we could be with Leo instead of that sadist.

Master Leo ordered me to open my mouth. He took my cheeks gently and looked inside demanding I lift my tongue. I was confused by this examination but did what he told me; grateful he was touching me anywhere. He seemed satisfied that the pills were not hidden among my teeth and smiled.

"That is my love. Okay Claus, he is good to go. Please be gentle with this precious boy. He is not well, and seizures could come without much warning. I am here by my phone if you need anything just call." He took my leash and handed it to Master Claus that had stood up ready to leave.

Master Leo leaned over and kissed my forehead then took a deep breath as Master Claus gently tugged on my leash. "Time to go, Mad Maxx. Follow me my love." The Elder smiled with a friendly look then opened the door and pulled me away from my lover Leo into the Haus hallway.

Master Bladrick followed behind me. I walked behind Master Claus feeling my heart falling. I was despairing over leaving Leo behind, but it was time for me to get back to work. I hated to admit that Mad Max was right. If we wanted to have a choice in who was touching us or where we stayed, then that fucking bat collar had to go.

Mad Maxx patted my back with a sigh. "This will be okay, Maximillian. We get to see Leo every four hours, we get three days not just once a week, and tonight he will be in the Great Hall with us. We work now and play later, ja?"

I frowned. "You mean we work now and Mad Max plays later. Der Hund says he gets to attend Leo, not us. We get the other rude bastards, brother. This is not fair, I tell you. I am speaking to our core about this. I want some time with Leo too if I must endure this other foul shit. If I must handle their coupling, then why not Leo's too? That doesn't make sense."

Mad Maxx shook his head. "Beats me brother. All you can do is ask Der Hund. I only know that we do what we are told, or we end up like Christian eventually will or Max before him. I know you think shattering is not for good because we see Max has returned. However, that Max attached to Mad Max is not the same shard. Look at how Christian fears him. He managed to keep all three of us from entering the flesh too. That is not a normal shard right there. He is something else entirely, but I don't know what. All I do know is that once you're gone you never come back. I wouldn't go pissing Der Hund off if I were you. Ask nicely and if he says nein, I will drop it."

I nodded. "Good points, brother. Still, I am going to ask him. That surely wouldn't make him angry." I was so busy deep in my discussion with Mad Maxx I almost ran over Master Claus when he stopped to open his apartment door.

To my surprise I saw Master Jonas approaching and moving fast I may add. "Wait Claus, stop. I need to see the boy a minute. I have his pills. He left them behind." He rushed up sweating, appearing out of breath.

Master Claus shot a look at Master Bladrick. "Jonas? What the hell is wrong with you scaring two old men by running like the loon. I thought the fucking Haus was on fire."

The Vampire snorted. "Not yet it is not. Here I need to give Mad Max his medication. I come inside with you and make sure he takes them."

Master Bladrick reached out his hand. "Give the pills to me, Jonas. We make sure he takes them. Give me the bottle too. I can fucking read the instructions you know."

Master Jonas frowned. "I would feel better if I saw him take them with my own eyes. No offence but the boy tends to cheek his meds."

I looked at the floor angered that everyone was calling me the deceitful one. I heard Master Claus scoff and Master Bladrick groan.

Then the Elder Claus said, "Enough, Jonas, I know how to check the boys mouth trust me. I have done it a few times, I dare say (he chuckled at that foul statement). Give me the fucking things and Bladrick and I will make sure he gets them. Now back off. Our clock has begun, and you have interfered to the point of making me more than a little angry."

Master Jonas rolled his eyes but gave the pills and bottle to Master Bladrick. "Just make sure he takes them every four hours." He turned then stormed off swearing under his breath.

I saw Master Claus shoot a look at Master Bladrick who returned it. The look conveyed that Jonas is kind of a sonofabitch you know. The two of them hauled me inside Master Claus's apartment and slammed the door behind them. The Elder released my leash went to the closest trash can and threw the bottle and pills inside. I gasped at that wondering what the hell.

Master Bladrick saw my look. "Leo suspects that Jonas is tampering with your medications, Mad Maxx. Claus and I agree. We got the Haus doctor to give Leo another prescription for the correct ones. You noticed the pills you took this morning were white. These Jonas gave us are pink. I don't know what he is pulling but we don't like it."

I held my breath in excitement. "Then I am not the schizophrenic, Master?" I was right that Vampire was fucking with my head.

Master Claus frowned. "Oh, you are indeed the schizophrenic, Mad Maxx. I am sorry love, you misunderstand. You have not been getting his pills for over two days now and you symptoms are worse than ever. We think Jonas was adding in something that fits into his scary beliefs about the Priceless powers and maybe an herbal block for your own ability to feel passion. He desires you never find any release until you take a mate. That is what that fucking chastity device is all about. Well fuck him. Bladrick, you got that hex wrench I told you to get? If so, take that Gott damned thing off the boy. It is ridiculous, this one-sided shit I expected from that sadist Peter but from Jonas. That I cannot take."

I almost died when Master Bladrick come at me with a hex wrench telling me to drop my breeches. I didn't argue but trembled while his old liver spotted hands removed it with skill. In only moments I was free of my chastity device once more. I began to believe all three of them wanted to trap me in my metal for good. What the hell was going on?

Master Bladrick must have seen my fear. "Relax my love. There is no way to know a penetration virginity has been breached in a boy. Only the words of an Elder or Voting Council member who witnesses the act can destroy your maidenhead. Claus, Leo, and I all agree that you have more than earned your rights to fair release during couplings. All of us are tops and pure schwulers. Therefore, you won't be penetrating a fucking thing during any intercourse with any of us. We don't view manual stimulation nor the blow job as intercourse love no matter what that old, tired law says. If you have never been inside the male or females sexual organs, you are the virgin, period. Jonas believes this too, but that fucker has a messed-up vision of this Priceless power of regeneration. He keeps you chased to prevent us from enjoying you and us allowing a payment for your suffering it. That can end right this minute if you allow us to honor the boy, we all love this way. You must vow to never tell Jonas of our mercy or anyone else for that matter. What do you say? We know you are a man of your word. Say it and receive equal service for the service you grant to us."

I frowned. "I vow I never say anything, but Master forgive me when I say you cannot release me. I don't find interest in the man."

He chuckled. "That is what Leo told us. That you are straight. So be it. You have your own hand and an imagination though, don't you? At least in our Haus and with Leo you can find your own thrill. We even get you anything you request to aid that. Magazines? Films? Such things exist, my boy. If you need a fantasy girl your Masters will supply them. Long as you never get them in the flesh, we turn a blind eye to what you do in your private moments to aid you to endure what you must. ja? All you need to do in return is keep making us the happy men you already have made us." He leaned down and kissed my forehead.

I nodded. "I thank you for the mercy, Masters. I am forever in your debt for it." Well, at least they allowed my masturbation which was better than Peter (I had to sneak around with him) or Jonas (that would have a shit fit over such things).

Master Claus laughed then sat down on his couch. "That is settled. Now Mad Maxx, I do believe awhile back you come begging me to take you while on your knees right in this very spot. I told you another day I would ravish you with vigor but had to deny that beautiful request on that night. I command you to come over here and let me prove my honesty. Bladrick you're welcome to join in whenever you are ready brother. I am thrilled to know you can gain your release through this marvelous collar's skills.

I am not opposed to sharing my tryst with him." I winced when I saw Master Bladrick nodding excitedly with a wicked smile. Mad Max was right. That motherfucker. I was going to play my Maximillian the London Bridge.

I endured this foul threesome action with the Elder schwulers without shedding a single tear, though I admit I almost broke down a couple times. I thought of Mad Max and the loving Leo many times to keep myself from coming unhinged. I wondered what dancing with him would be like and imagined being at the Straußenfarm with him. It helped get me through it until Master Bladrick released his nasty bloody urine into my mouth.

Master Claus was not done with his lustful intercourse yet. I was stuck unable to spit this shit out without soiling Master Bladrick's lap. With much disgust I just swallowed that crap. Holding it in my mouth till Master Claus found his climax with me wasn't an option.

I thought for sure I would vomit but the Elder's harsh thrusting into my still very sore personal spot kept me distracted from giving Master Bladrick a Roman shower. *Thank Gott. Okay that is a tough call there. Not sure which was worse? It was bad, that is all I will say about it. I would say I had endured worse, but this was sure up there in the top five of horrific couplings Maximillian ever can claim to have been forced into.*

I nearly let out a sigh of relief when I heard Master Claus groan out in his apex. All I could think of was a hot shower and maybe a purge too. To my despair that luxury would be put off awhile longer. The two of them wanted to cuddle a bit after their defiling of the boy. I was near madness as I tolerated more of their wanton touching and probing of my Haus. It seemed to never end.

Then to make it worse, they took turns grabbing my cock. I had to endure this horrible business – their probing was painful as hell for the record – as they attempted to get me to demonstrate interest in their molestations.

Finally Master Bladrick called off their tag teaming. "Give it up, Claus. The boy is straight just like Leo said. That really is too bad. Well Mad Maxx I must say that is rough for you since your Masters need a bottom and you are the top, and not a schwuler to boot. It shows that Gott has a twisted since of humor." *I found nothing funny about this shit. If there was a Gott, and there is not, he was not someone I ever wanted to know. This life I lived was beyond humiliating.*

I was at last released from their special service request. I practically ran to the washroom scrubbing off another couple of layers of my flesh over that ordeal. I was brushing my teeth, for the third time in a row, when Master Claus barged in through the door.

He looked at me with lust still in his expression, thank Gott I was dressed already. "Mad Maxx your artistic skill has worn my brother and me out. You are quite the dynamo, and we are old men. We wish to nap for a while. Leo says you have not been eating. I command you to go downstairs and get your lunch in the kitchen. Then tell

Evelynn to give you the sack of dry bread for Claus and Bladrick. When you finish getting your medications from Leo in three hours your Masters will take you for a treat. You have earned it and more. We like to feed the ducks at the pond in the yard. Today, we take our Priceless collar with us for fresh air and sun, ja?"

I nodded. "As you wish Master." I was damned glad that not only would they be literally off my ass for the next several hours but to get to go outside was indeed a wonderful gift. Though not worth what I had to pay for it.

I rushed to finish my hygiene then practically fled their apartment grateful to be free of them for a bit. I started to head for the back steps when I recalled I forgot my coat and sunglasses. I started to head back to get them when I saw Master Jonas coming. I was out of my chastity device and sure as shit didn't want any more pawing at me. He had not spotted me, so I rushed to hide around the corner holding my breath.

I watched him pass humming a tune as he went down the steps slowly. I realized if he looked back, he would see me standing dare. I took off the way I had come headed back towards Master Leo's apartment. I thought maybe he could loan me another pair of glasses, okay I admit it, I was just going to see him to see him. There are you happy now Mad Max. I admitted it.

Anyway, I was almost to his door when Mad Max and Max swiftly knocked me and Mad Maxx to the floor. I rolled along the hallway pissed out of my mind. That bastard was hell bent to keep me from enjoying any adoration from that desirable Leo.

He smiled at me with a look of humor. "Shame on you, Maximillian. You heard Der Hund. I was ordered to take back the wheel the second the couple was over. I gave you time for the shower and teeth but now I will do my job. Thanks for thinking what I would have thought though. Smart move to come see Leo." He winked at me.

I growled out while shooting a look of fury at Mad Maxx that was now standing up also angered. "You could have shown a little mercy brother. That shit with Master Bladrick and Master Claus was awful. I could use some real affection to make up for it. Let me and Mad Maxx take the wheel just a bit longer?"

Mad Max laughed. "You ask the sadist for mercy, Maximillian? I think not. Go cry if you want but I care nothing for your hurt feelings. What, wait, huh, okay. Fuck Maximillian, Max says mercy granted motherfucker. This is bullshit." I watched in disbelief as Mad Max and Max disengaged the boy and Mad Maxx, and I were swept back inside.

I stood their staring at the irritated Mad Max and Max. "Uhm, thank you for the mercy brother." I couldn't believe he relented.

He crossed his arms with a look of fury in his eyes. "Thank Max. He says I need to learn to fucking share the joys as much as I divide out the burdens. I want to kill all three of you cocksuckers right this minute, but he is speaking wisdom. Go enjoy the adoration but if it gets heated, I am kicking your ass out Maximillian. Leo is mine."

I nodded with a smile. "You got it brother. I only want to dance and be held by him anyway. I have had enough of that penetration bullshit today, trust me."

He scoffed. "I don't trust you. That is why I stick around to keep my eyes on you, whore." Well, what did you expect? It is the Mad Max after all.

I turned to head back toward Master Leo's apartment with a happy heart. I couldn't wait to try his dancing out and just be adored without the sex crap. I walked by the dead Barnim's door and felt a chill go up my spine. My back was still not healed from his horror. I walked over to the banister and looked below. I could see the illusion of him lying there in his blood. I shivered then turned around.

The stinging chemical flew into my eyes and mouth sending me to my knees immediately. The rapist was attacking me on the Elders' fucking floor in broad daylight with Master Leo only feet away. I couldn't believe this shit. I struggled to breathe, choking, and clawing at my face.

I felt the man grab my shoulder dragging my weight without trouble. I slapped him and tried to scratch his hands that pulled me along. Suddenly he stopped then grabbed me throwing me over his shoulder like the sack of grain. I kicked and pounded on him in fury, completely blind and barely able to breathe as I coughed uncontrollably.

I heard him unlocking a door then he went inside. I knew the smell immediately. This was Barnim's apartment we had entered. That foul place's scent was branded in my mind after what he had done. I whimpered when I heard him close and lock the door. I couldn't scream nor beat his strength. I felt horrid despair fill me as he threw me to the ground. I couldn't handle another forced intercourse. I just couldn't do this. I felt my own real sobbing joining the buckets of pepper induced tears.

Then suddenly I was thrown free with my brother Mad Maxx. I still couldn't see a fucking thing, but the flesh was gone. I was in the nothing. I called out for Mad Maxx.

"What has happened brother?" I wailed thinking the boy had been killed (had to be.)

I heard him call back. "I don't believe it but just before we were tossed, I saw Mad Max and Max take the wheel brother. He has taken the task of enduring the rape and beating for us."

I couldn't believe my ears. Mad Max showed me mercy a second time. What the hell was going on? I thought Mad Max hated me. I sure hated him but now I had to admit I suddenly respected him. After all, he is the lover of us all.

Chapter 45: An Inch of Flesh for a Bloody Pound

Max had pulled me into the boy when that pepper spray flew into Maximillian's face. I couldn't believe this crazy shit. The flesh was about to get raped, then the hell beaten out of it. This dumb motherfucker made us leave when Maximillian was on the way to see Leo but decided to enter now at this moment of horror. What the fuck.

I glared at Max with hate. "Why? This situation is hopeless Max. The boy is about to be forced into a brutal couple and there is nothing we can do about this. I don't want to deal with this shit."

Max looked saddened. "Brother, Maximillian doesn't have the strength for this dishonor. We do. Brace yourself. That criminal is standing there surveying the flesh. He is about to start his assault. I will hold your hand if you need me too. This will be worse than the last two. This bastard is getting more violent with each attack. He plans to really hurt us this time."

I growled. "Nein, I don't need to hold your fucking hand, schwuler freak. You can suck my cock though. Idiot, this is bullshit. Why the fuck must I be stuck with you?"

Max shook his head. "There is a price for everything Mad Max. I am yours."

I shook my head in fury as I felt the rapist lifting us up to drag the boy into the back of the apartment. "Oh? I pay for what? Being the hostage of this fucking Haus and collar."

He looked at the floor. "For being born, brother. I am the soul. You are my shield, Mad Max. You are the strongest of all the shards for that very reason. Now do your job. Protect me from the outside and keep me untainted by those that try to crush me. You have no empathy nor compassion for the ones that have none for us. That allows you to hear clearly and see through most lies. Harsh words cannot harm you, and betrayal is expected so that too never sticks to your tough skin. In fact, only actions that you don't expect from another can cause you to become confused, unsure how to proceed. I could see Leo clearly and Annette because of that very behavior. The expression of true love from the soul of those around you can move you to do amazing kindness without expectation of return. Maximillian and Max Maxx do their jobs with broken heart, but you my sadist have an unbreakable one. That is why you were that shard that gave back Der Hund his lost soul. Brother, look out, he is coming." I winced as Max came forward and attached his boundaries to my own. We became one.

The rapist tore off my breeches near ripping them. I attempted kicking at him with all my might. He grabbed my right leg and punched me hard in the thigh. I gasped out in agony believing he broke the Gott damned thing. For the moment I was stunned to stupid from the pain.

He roughly pulled me forward then forced his entry, spitting like he always did. I groaned out as he began his fast-brutal thrusting. I reached up scratching for his eyes hoping to blind him as he had me. I brushed his nose and without hesitation I reared back and punched him in it. He stopped his thrust but didn't uncouple with me.

I felt him tremble as he fell forward wailing out loud in a high-pitched sound. I listened intently. This was not a noise from anyone I recognized but then again, a scream is not the same sound as a normal pitched voice.

That rapist short punched me in the chest. I felt the air rush from my lungs. My hands released as I tried to cover my torso reflexively to prevent another blow. He backhanded my face with much force. I felt my teeth rattling in my jaw as he landed another blow in my stomach. This effectively ended my struggle for several minutes. I fought to breathe, nearly unconscious as he resumed his brutal raping of the boy.

I lay there enduring his penetration thinking I would die this time. No matter how hard I tried the air wouldn't come in well. I was weak from the lack of it. The man withdrew and spun me to my face. He re-entered and began his intercourse anew. I gasped like a fish out of water, unable to do anything about this horror.

He took much time reaching his orgasm. Longer than the last time in fact. When he got close, he would reposition me and start over again. I was in and out of consciousness throughout most of it. I wasn't getting enough oxygen for some reason to remain awake enough to fight with any vigor.

Then finally he decided he had violated me to his satisfaction. He cum with a loud moan as always but no other sounds but his panting. I kept my burning eyes closed. I couldn't see anything with them anyway. I couldn't smell neither after the first few moments of that fucking spray. If that sonofabitch wore cologne or had a favorite soap, I wouldn't know it.

He withdrew and left me for a moment. I thought wildly of the possible identity of this cocksucker. How could he know I was alone on the Elders' floor? Where did he get those keys? Was it possible this was my Master Jonas? I wondered if he had been following me. I thought he had not seen me hiding when he went by but maybe he did.

I thought of the other two rapes. Maybe he found out about the blow job to Leo in the torture chamber that day. Then the next one, maybe he knew about my seduction of the H twins. That had to be the answer.

I knew he was the jealous type and he had been tampering with my medications. Maybe he wanted to put me out of the rotation by beating me so badly I would need a rest. This was making more and more sense to me.

He would have access to Barnim's keys, had time to follow me too. He would know about the secret places in the Haus and could track my coming and going. I grew quite angry that bastard was raping when it was not necessary. I'd do whatever the hell he wanted without him having to beat the stuffing out of me.

Gott damn this sonofabitch; Jonas was my assailant. I wished Christian was around. He'd show that Vampire a stake to his fucking heart. If this was his idea of loving me then he could shove that shit right up his ass instead of mine.

I felt the rapist putting back on my breeches. He stopped briefly when buttoning them. I was still helpless to even lift my arms. My breathing had not gotten any better. I trembled wondering what he was doing. Then I felt him grab my cock. I gasped in terror that he planned to rip it off.

Instead, he appeared to be examining it. What the hell was with this guy? He did this for several moments then dropped it and finished dressing me. This man acted as if he were curious to see manhood. As if my sex was something he never seen before. Wait, I had always been in the chastity device. He hadn't been able to view it before.

I reached deep inside myself and found the strength to set up and grab blindly in the direction of the sounds of his breathing. I got lucky and managed to snatch his hair. I attacked with both hands tearing at it and noticing it was short. I did my best to get a handful to see the color when my vision cleared, if I lived that was. He easily knocked me off him. I had not strength to hold on.

This immediately ruled out the Vampire. The Elder had long hair like the frau, and he had seen my cock many times. If it was not Master Jonas, then who the fuck was this motherfucker.

I felt the man run his hand along my swollen cheek. I turned my head trying to avoid his touching. He grabbed my hair and lifted my grasping head. I could feel his breath on me. I lifted my weak arms and tried to push him away. He shook me hard, nearly sending me back into the void. My hands fell uselessly to my sides as I fought to stay conscious.

The rapist got back into my face then turned my head and licked into my ear. I gasped and coughed harder from the sickness rising inside me. He then released me. I fell hard to the floor. I rolled over and decided to try to crawl away. The man allowed this making sounds under his breath that sounded like a whispered chuckle.

I must have looked very pathetic. I would pull up on all fours taking a few steps slowly on my knees then fall face first in coughing fits. I finally gave up and rolled into a ball unable to continue without rest. I thought for sure I was dying. The struggle not to smother was now scaring me a great deal.

The rapist came at me and lifted me by my shirt like a rag doll. I quivered and coughed as he held me there for a moment. Then he flung me across the room. I hit a wall hard, falling unable to catch myself. He was on me in seconds. With catlike speed he threw me back the other way.

I again hit a wall and fell helplessly to the ground. I was nearing unconsciousness when I felt him lift me again.

This time he backhanded me as he held me up. I was tossed backward onto a solid surface that I thought was a tabletop.

He kneed me in my hodensack sending me right to hell. I reached down to hold my sensitive parts. The man grabbed my left wrist pulling it from my cradling of my manhood. I let out a raspy yell as the rapist grabbed my ring finger on that hand and broke it with his bare hands.

He then spun me to my face bashing my forehead on that surface of the table till I nearly couldn't remember my own name. He finally stopped when I went limp losing consciousness at last. I was barely aware of him pulling down my breeches to go for seconds. His harsh entry was the last thing I felt before I drifted off into the void of nothing.

I awoke in a bed groggy and in much pain. I couldn't move a thing. I groaned out loud grateful I could breathe. The act of taking in air was beyond painful in my chest. I thought I had been stabbed for sure. I am unable to gain my location.

The face of Master Leo appeared above me looking frightened to death. "Christian Axel? Can you hear me?"

I nodded but winced then coughed out. "How did I get here? What has happened? I cannot breathe."

Master Leo gasped but smiled. "Oh, thank Gott. I thought for sure we lost you, my love. You have a broken rib, collar bone and finger. When you didn't show up for your meds, I came out of my apartment to find you. I found you alright. Laid out unconscious in front of my door. Do you remember anything? The doctor thinks that you may have had another seizure?"

I suddenly recalled that brutal raping shit and realized the Haus doctor had not discovered it. thank Gott.

I nodded. "That is what must have happened, Master." I began coughing with much pain in my chest, hell everywhere.

That rat bastard rapist had finished his fun with me, redressed me, then dumped me in front of Master Leo's door. It was becoming clear that his plan was to get me to think Master Leo was the one doing this shit. That motherfucker.

I heard the voice of Master Jonas call out. "Leo, how is the boy? Is he awake yet?"

My Master rolled his eyes at me then turned towards Master Jonas's voice. "Ja. He just regained consciousness. He says a seizure, as the doctor suspected. He will need rest and quiet Jonas. I say we don't move him for a few weeks till his ribs heal."

Master Jonas growled with fury. "Fuck that. He will be coming home with his man, Leo. Every fucking time I let him out of my sight this shit happens. Claus, Bladrick and you are fools. This Priceless is doomed to death if I keep letting you all leash him."

I heard that shit. I couldn't let Master Jonas piss off his brothers in his attempt to keep me from their rotation rights. I was trying to unite the group of Elders, not bust them apart.

I attempted to sit up doing my best to ignore the agony of that. I noticed my chest was wrapped up in a tight bandage. That pressure from it was making my breathing shallow, not the broken rib. My left shoulder and left hand were also in dressings.

I forced a deep breath wincing with pins and needles in my chest. "Master, please this was not Master Leo's nor Master Claus's fault. I was going to my lunch when I see the auras of the Grand Mal. I came running for Master Leo's apartment because it was close. I didn't make it, I guess. I am okay now; I can get back to my rotation. The pain is minimal." *Now there was one hell of a lie. Yikes!*

Master Leo looked at me with a startle and tried to gently push me back down into the bed. "Mad Maxx. You are ill. Lay down. This is not your business, love. I let this insolence slide because is due to head injury, ja?"

Master Jonas laughed with wickedness. "The boy says he is fine mother hen. My Mad Maxx is the Priceless. He can overcome anything. This was a little spill is all. Let him speak for himself Leo. My love, tell your Master Leo I know what I am about."

I nodded, doing my best not to pant from the pain. "Ja, ja Master. Master Jonas is right. I am fine. I can attend my services. I need no rest. How long was I out?"

Master Leo scoffed. "This is bullshit, Jonas. This boy is mentally ill. The schizophrenics do not mind their injuries like they should. He will hurt himself trying to please your stupid ass. All you have brainwashed this poor kid. I won't stand for it. You go get Claus and Bladrick. I want a vote on this. Mad Maxx needs rest, Gott damn it."

Master Jonas chuckled hard. "You were only out a few hours, Mad Maxx. It is not even dinner time. I go get the brother Elders snoozing on your living room furniture for this vote. I warn you Leo. This will not go your way. I will have my man home midnight tomorrow." He left the room to retrieve the Elders.

Master Leo shot a look of fear at me. "My love, what really happened? This is far too much damage for a simple seizure. You have been beaten to shit. The others may buy this, but Leo is no fool. You are covered in black from head to toe. Some of the marks on your shoulders are in the shape of fingers from hands. I noticed that these injuries have been steadily getting worse over the last four days. You can tell your Leo the truth about it, my love. Is one of your Masters hurting you? Is it Jonas?"

I wanted to tell Leo, Meine Liebe, I really did. I couldn't though. The rapist was likely on the Voting Council. That meant he would not really be tried for this crime, but I was sure as shit would be. The second the Council heard I was raped they would claim a theft of "Priceless proportions" had been committed against my Elder Masters.

Since in the Haus, as the forbidden silver, I was viewed as a seducer of unearthly abilities. There was a widespread belief that one couldn't blame any Dominant for being so tempted to taste my metal. The Council would view this crime as my fault because I knew my special services are prized so highly.

Therefore, I should have protected my Masters' rights better. There was no way I wouldn't be held as the instigator that asked for these foul assaults.

Once found guilty, and I would be trust me, a penalty was lifetime collaring and auction. My silver would then be sold to the highest bidder, and likely I could expect to be passed around the Haus.

In short order, I would be torn to bits in the feeding frenzy of those perverted sharks that called themselves Dominants.

Even my powerful Masters couldn't help me with this one. The voting Council could and would take possession of my collar from them without allowing trial or argument. I was in a lot of trouble and the last person on Earth I wanted to take down with me was my Leo.

I had to find this man and kill him without getting caught. I knew Leo would want to help. If I got caught, I would be put to the yard, no questions asked. If Leo helped, they would exile him forever or worse send him to the orchard with me.

Whoever was doing this wanted me to either believe Master Leo was the rapist or involve him in the knowledge of the crime. I had to refrain from giving that motherfucker his wish. No matter how scared I was, this guy eventually was going to kill me. I would have to deal with this in silence.

I now realized that somehow Leo was tied to this horror show. I was thinking the rapist must hate him or hold a grudge for some slight. I thought it likely the man intended to get his pleasures from me until he tired of them. Then, eventually, get me trapped in my metal and sent to the auction. This was sure to hurt Master Leo and all the Elders at the same time. You know, kill many birds with a single stone, ja?

I thought of the brutes Grisham and Malfred from the confrontation at the club. I recalled that Grisham had threatened Leo. He brazenly stated he wanted to rape me to nothing in front of my Master after beating him to weakness.

It seemed to me; this cocksucker Grisham had to be my rapist. He matched the strength, size, and he was on the Voting Council. I notified Max to let Der Hund know the second the boy was able, we had to weed out Grisham as the most likely culprit of the rapes. *I prayed our core had not shattered our brother Christian. We never needed him more in our life.*

I shook my head keeping my eyes locked onto his. "Nein, Master, I merely been having violent seizures. I understand that now. I was skipping my medication like you suspected. I pay for this stupidity with injury. I swear to you it won't happen again. Forgive me for my lack of insight?" I felt terrible lying. I simply loved him too much to risk his life. I was aware it was highly likely my attempts to take out this powerful Dominant would fail.

My Master sighed with much sadness, then kissed my forehead. "There is nothing to forgive my love. Fear of the pills is part of your disease. You cannot help it. I will help you beat that terror. That said, I want you to lay down. These injuries you suffer are severe. You need rest, time, and peace. I was terrified I lost you this time. Maybe next time I don't get so lucky, nor do you." I allowed him to press me down to the mattress to my back without a fight.

The Elders came into the room like the gangbusters, all three of them yelling at each other. I laid there doing my best to brace against the agony. I really needed to convince them I was fine. No matter what, I had to get out of the bed right away. I had a murder to plan.

I sat up once more when Leo turned his attention to the grumbling old men. "What the hell, brothers. What is all this noise? Mad Maxx needs quiet. Hush this discord at once or at least take it to another room." He said while grabbing his chest as if horrified at their behaviors.

Master Claus saw me sitting there taking shallow breaths. "The collar looks like death warmed over Jonas. This boy needs rest. He is safe here with Leo. What the hell is your issue with it? When he is better, we can move him then."

Master Bladrick grumbled. "He does seem blacker and bluer than I have ever seen on anything less than a corpse. Jonas, leave the boy alone. Leo can take fine care of him. I agree with Claus."

Master Jonas growled out. "He can fucking walk. He is my man. I want my blood bonded home with me. I can take of my own without the aid of fucking Leo."

Master Leo put his hands on his hips, appearing startled. "Oh, I am now called by my brother "Fucking Leo?" Really, Jonas? What the hell is wrong with him staying here with me? What exactly do you fear will happen here in my Haus? I want to hear this."

Master Jonas glared at him. "Who the fuck knows what you do, Leo. You run around this Haus stirring the pot wherever you go. Everyone knows you're a fucking weasel. A gossiping Frau is all you are. How are Peter and Gretta these days? They are your best friends, aren't they? You know damned well they are seeking this collar's blood and yet you still get on your knees to suck Peters' cock. No telling what a man without loyalty would do to my precious Christian Axel. I noticed ever since you joined us on this floor my Priceless has started to look like a car wreck victim."

Master Leo scoffed then let out a yelp while snapping his fingers into the air (he is a queen, remember. "That is not fucking truth and you know this, Jonas. I left the side of

Peter and Gretta a long time ago. The only reason you all hate me is you know damned well I report the abuses to the silvers without hesitation nor care for who is the culprit. This Haus is full of the foulest of criminals and I at least do something about it. I don't care if they are new Dominant or Elder, you hurt a silver or break the rules I am going right to the Council with it. If this pisses everyone off too bad for them. If they treat their collars like human beings, they have nothing to fear from Leo." He snorted and stuck his head into the air with much drama.

Master Claus groaned at that over-the-top effeminate behavior Leo was demonstrating. "Stop that righteous indignity bullshit, Leo. You have indeed turned in almost every fucking Dominant in this Haus with cases claiming they broke the rules with their collars. I for one don't care, but many around here have suffered heavy fines for it. You made no friends when you got Beatrice exiled. You have earned your reputation fairly I would say. However, Jonas, the way he is viewed by the members of this Haus has nothing to do with his capability to nurse our Priceless back to health." He glared at the Vampire with irritation.

Master Jonas scoffed "The hell it doesn't. I tell you he is trying to get the Priceless collar back into the clutches of Peter and his group's control. He is nothing but a mole for them. I have it on good authority that the other they he was spotted in a friendly discussion with the head of the Voting Council, Gretta. They were all pals again, thick as mud. You head my warnings brothers. Leo is here to trip up our collar and trap him in his metal."

I gasped at that and shot a look of fear at Master Leo. He looked back at me and winked. Then turned around and let those old schwulers have it with both barrels.

"I sure as shit was talking to that old bag Gretta. I don't care who the fuck knows it. I ran across her in the dungeons while walking the boy and reminiscing about my own days as the Dungeon Master. She came up to us and I spent some time trying to pick her clean of any future moves of that trio of scum. The bitch fell for my pretending all was well, but sadly I could get no information from her. Wait, nein that is not that truth. She did tell me you Jonas sold this boy's penetration virginity to her. Humm, now that is not nice, is it? The boy was used by those monsters and near sent to the yard, but you love him so much you sold him to Gretta's nasty bed. Disgusting. As it is, ask anyone around that Dungeon, and there were many, what happened when I was done using that whore for my intel. I told her where to get off and how to fucking do it loud enough for everyone to hear it. I was so rude, even if I had been acting Gretta could not allow it to slide. Besides if I were trying to hurt Mad Maxx, tell me what the hell would I be getting from it? Am I not already an Elder? Am I not already wealthy? Do I not already have my own private taste of the most valued metal in this Haus? You tell me what Peter and his hooligans can offer to Leo that I don't already have." He crossed his arms and flounced like the frau.

Master Claus chuckled with his partner Master Bladrick at Master Leo's most accurate statement. "He got you there, Jonas. Look you are outvoted. The doctor says there will be no strenuous activity for six weeks. The boy stays here until his pain is under control. Then I will decide how to proceed from there. That is my judgement. You are out voted brother Jonas."

Master Jonas bellowed in fury. "This is bullshit. I am not putting up with it. This collar is my blood bonded. I claim my rights outrank your ruling."

I took a deep breath and braced for a backhand while speaking without being asked. This is serious insolence right there, but I had to stop this before it got out of hand. "Master Claus, Master Bladrick, may I speak to you privately for a moment? I ask the mercy with only the greatest of respect and demand the punishment for daring to interrupt my betters in their discussion."

All the Masters stopped arguing and stared at me in surprise. This was not usual behavior for a well-seasoned submissive. Their shock at my daring to incite severe anger kept them silent in the room for several minutes.

Then Master Claus found his tongue., "I am going to grant you this request Mad Maxx. I do it only because I want to see what the hell you need to say that you are willing to get thudded in your fucking condition to ask it. Leo and Jonas get out. I will speak to this collar, then decide on his punishment for insulting us." I winced at those words but watched an irritated Leo and very pissed Master Jonas leave immediately.

Master Claus and Master Bladrick approached my bed. Master Bladrick appeared humored, but the crossdressing Elder looked furious.

"What the fuck, Mad Maxx. Now I have to beat you. Start talking before I know you the fuck out for embarrassing me like this in front of my brothers," he growled out with brimstone in his eyes.

Maser Bladrick put his wizened hand on Master Claus's shoulder. "Calm down old man. You are scaring the boy. Let him say what he needs to say before you go threatening to beat up one that even I could lick in the state he is in." He chuckled at that.

I looked at the floor. "Thank you for your mercy, Masters. I beg you to allow me to stay on the rotation. I am injured but the seriousness is cosmetic only. I can survive on pain killers. In a few days this will pass. I am capable of service. If you pull me from the rotation Master Jonas will become despondent, I fear. I think that would not be helpful to gain his allegiance to the Elder brotherhood. I say that width respect and accept your punishment as you see fit for daring to be insolent."

Master Claus snorted. "Jonas needs to learn his fucking place, Mad Maxx. I refuse to have this discussion with a submissive. I am the second in command in this Haus, but everyone keeps running me over, Gott damn it. Even this kid thinks he knows better than Claus."

I shook my head. "I beg your forgiveness for my poor communication skills, Master. You are correct. I have no ability to speak to a Dominant at their level. I would restate my request in a simple term fit for one in my lowly station. Master Jonas is critical to my breaking the collar. If you allow the rotation to continue as it is, I will offer to come to you for payment if you grant this favor."

Master Bladrick laughed hard at that while Master Claus shot him an angry look. "What is this you say, Maxx? What the fuck can you offer your Master in payment? I own your collar. I take what I want when I like."

I nodded. "This is true, Master, and I will serve you to my truest ability always. However, if you enjoyed your time with me today, both of you understand I can make that change without denying your orders. No matter how much I do adore you two, I can offer resistance to your orders with a dark attitude without any emotion in my service provisions. You may punish me for saying that, or for withdrawing my affections, but I will not relent. Without Master Jonas on my side, I am doomed anyway. I would rather you beat me to death then to be trapped in my metal by Mistress Cora. Master Bladrick, you know laying in a bed slows the healing." I was pushing my luck, but I had to make this Elder listen to me.

Master Bladrick coughed and smiled wide. "Give the boy what he wants. I like his style. He is also right. He is a sexual artist most rare, Claus. I for one don't want to give this thrill we had with him up. If he wanted to be the dead fuck it would rapidly become a buzz kill. He demonstrates his inner strength this moment. If he says he can keep the rotation, I believe him."

Master Claus's expression had softened upon hearing my threat, "You are ill, Maxx. I don't mean to be cruel to you, but rest is what you need. You won't break that collar if you die trying to serve us either. There must be a compromise."

Master Bladrick put his finger to his lips pursing them. "Ah, I have it. We have all this day and tomorrow till midnight with the boy, brother. Let's leave him here and allow the pain killers and muscle relaxers to work while he rests, ja? Then Jonas can surely move him to his own bed after that. It is only a few broken bones. None of them keep him from getting around. Jonas will be happy and so will Leo. The boy can make up the lost day to us during Leo's next rotation if he agrees this is fair?"

I nodded with a smile. "Ja, I agree to this idea Master. As always, I bow to your incredible wisdom."

Master Bladrick scoffed with a cat smile. "Save running that silver tongue across my hodensack for when I can truly thrill at it Maxx. Claus, come on. This will work. Trust me on this."

Claus appeared to think deeply for a moment then sighed. "Okay, I agree to this plan as well. I suppose if we do this it is not really giving in to Jonas nor to Leo. I am still the boss, ja?" Master Bladrick and I nodded as Master Claus clapped his hands demanding the rest of the Elders be allowed to re-enter the room.

Master Jonas and Master Leo entered the room when Master Bladrick called them. They were shooting "go to hell looks" toward each other. Their aggressions were more than a little evident. Master Bladrick cleared his throat to get them to focus on Master Claus.

"I have decided that Mad Maxx is well enough to finish his rotation for this cycle. Jonas, you may come pick him up from my apartment tomorrow at midnight. My ruling is final." Master Leo immediately crossed his arms with a look of disbelief while Master Jonas grinned his toothy smile of victory.

Master Jonas then spoke. "I am very happy that is settled. I will see everyone in a few hours at the Great Hall." He turned and stormed out of the room with speed.

Master Claus and Master Bladrick shot me a look of concern. Master Leo stomped his foot and flounced.

"I expected this kind of barbarous behavior out of Jonas, but you two. Shame on you. Your wooden heads must be full of fucking termites by now. You cannot control your urges by this age then there is no hope for the world." He scoffed out.

Master Claus shook his head. "You are speaking without knowing shit, Leo. The boy is not going to leave this bed till midnight tomorrow. He won't be attending our fucking lust nor does our wood have bugs. I can take that boy like the young stallion even at my advanced age. Ask him, I don't appreciate that insult."

Master Leo's expression became confused "What? You say he stays in my bed till Jonas's turn? You are not going to haul him off for your services?"

Master Bladrick snorted. "Get a load of this prissy bitch. He says we are ancient, but he is the one going deaf. Ja Leo, which is what Claus says. You keep him still and stress free. Attend his injuries. The pain killers should keep the agony of them to a dull roar. Make sure he moves about as he feels strong enough to do. When eleven thirty tomorrow night comes, Claus and I will retrieve him. Jonas need not be the wiser, ja."

Leo dropped his arms and looked at his shoes. "Nein, he need not know of this. However, this boy will not be well in such a short time. When Jonas comes for him then, he will not mind his health like we can. The pills, his weird beliefs, they are dangerous."

Master Claus shook his head. "Look Leo, I give you three days in the cycle and the rights to adjust the collars medications. Bladrick and I give up our days in this cycle to entrust him to your care. This will have to do to settle your fears or get over them. We all have to share this boy's affections and artistic skills. There is no way to please all of us at the same time unless you want to move in together in the commune with orgies every night."

Master Bladrick looked at Master Claus with a huge smile. "Now that is the best fucking plan I have heard in ages. What do you say, Leo? Want to be my roomie? We can all share Mad Maxx equally." He started laughing with Master Claus joining in at the tease.

Neither Master Leo nor I thought that very funny I noticed. He stood there rolling his eyes then came rushing to the bed sitting down on the edge. He reached out to feel my forehead for fever. The other Elders spoke of their youth and wild conquests loudly behind Master Leo.

Master Leo shot me a look of disgust then yelled out politely. "If you gentlemen wish to have refreshments, I believe there is a bottle of whiskey behind the bar. Otherwise, I would like to beg you fellas to lower your noise for our little sick bird's comfort. Thank you for understanding."

The two Eldest Elders chuckled, then pretended to race each other for the door to get to the brown liquor.

Master Leo widened his eyes with humor at those buzzards taking off for the whiskey. "That takes care of them. After a thimbleful in their ancient livers, they will be snoring like the chainsaws, ja? I happen to know they are both cheap drunks. Good thing, too. Makes it easier to get them to believe they partied like the animals they wish they still were. You rest my love. I will go and undo their flies. Then later when they awake from their stupor, I will tell them we had that wild orgy. They won't recall it, but you and I will swear it happened. Those old fuckers will be upset for weeks that they missed the whole thing. That ought to teach them a lesson about talking like brutes, ja?"

He made me laugh by making silly faces but his threat to trick them was enough.

I groaned out in pain from my giggling. "Ah Leo, I don't think they will buy the rouse. I couldn't grant the special services to a flea right this moment."

He leaned down and kissed my lips lightly. "That is why I was trying to keep you out of Jonas's claws, my love. You are never going to heal mind or flesh at this rate. No one ever lets you, ja? What did you say to Claus to change his mind?" He looked at me with seriousness in his expression.

I looked away from his searching eyes. "My love, I am sorry, but I begged him to rescind his judgement for releasing me to rotation. I cannot have Jonas angry with you or the others. You don't know him like I do. If you fool with him, the results can be foul. I would beg you to let me handle the Vampire and stay out of this situation I find myself in with him."

Leo sighed. "Christian Axel, I only want the best for you. I love you too much to sit back and let that selfish bastard misuse you like this. He is only interested in some fantasy of eternal life. Mark my words, if he thought he could live for even five extra years by cutting your throat, he would do it without remorse nor hesitation."

I nodded, still unable to look into my lovers eyes. "I am aware of this Leo. I know you say I need not trust you, but that is not the truth. I love you and I know I do because I believe in your words to me. I ask you once more to allow me to deal with Jonas without your interference. Take the leap of faith with me on this one, my love. If you cannot do this, then our affair will fail and both of us will be broken hearted."

Master Leo gently pulled my face to stare into his own. "If you tell me this is your desire, then I do what you say. I can deny you nothing, my love. All I care is that you are safe and comfortable as possible given the horror you endure here. I pray that a better day will come for you. Then maybe you leave this nightmare world of being encased in silver. When that time comes, you will leave your Leo for good. I will grieve for myself and celebrate for my Christian Axel on that momentous day. Until then, you tell me what you need, and I will do whatever it takes to make it happen." He leaned in and kissed my busted lips with much tenderness.

I winced but enjoyed his attention until he pulled out of his adoration. "Thank you, Leo, for everything you do for me. I swear to you no matter where I go or what happens I will love you for always. My heart is not a fickle creature. It is the cold stone. You manage to get inside, there is no escaping it. You will never have cause to grieve over the loss of what can never be lost."

Master Leo closed his eyes with a beautiful smile on his face. "And just when I thought I couldn't love you more than I already do, you go and sing to my heart another perfect song. I want you to take these pain killers, your medication, and eat something, then rest. When you are well, I have a new record for you to listen to with me. I will wait to have my Christian Axel with me when I open the wrap. If I am the lucky bastard, maybe that beautiful boy will dance with me again."

I nodded. "Sounds like a reason to hurry up my mending. Is this one in English too?"

He nodded. "Ja, all the best disco and underground music is, my love, why?"

I smiled. "Can you get me some books to learn the language? I would like very much to know what my lover sings in my ear when we sway together."

Master Leo snort laughed. "Ah, well I can tell you I sing I love you repeatedly in every language." He stroked my swollen cheek lovingly, then said many other romantic things for several moments.

I enjoyed his pampering very much. I did take my medications without argument, but the food was a bit more of a struggle. That horrible metal taste bothered me. Leo insisted I eat the whole plate while he watched. I almost vomited twice. I noticed he didn't look happy at this behavior.

Leo told me this strange metallic flavor was a hallucination of taste that I was experiencing and not real. I wanted to believe him, but I knew Master Jonas. That clever Bat could be capable of anything, even tainting my food.

When I finished the meal he kissed me, then turned out the light insisting I get my rest. I fell asleep deeply thanks to the sedatives and pain killers. I barely awoke when later that night Leo quietly crept into the bed. I tried to move in closer to him when I felt his arm gently wrap around my lower chest in his cuddle. He whispered for me to go back to sleep which I couldn't help but do.

The next morning, I came alert to Leo insisting I take more medication and eat breakfast. I was groggy but followed his wishes without argument. I wasn't feeling well at all. The bruises and cuts from the attack the day before had deepened into ugly colors of dark blacks, purples, and reds. I looked like a boy with some strange disease of the flesh.

Leo would winced each time he looked at me and said, "Damn, it hurts me just looking at this. I cannot imagine how much pain you must be in, my love." I must have heard him say that a hundred times before I finally asked him to stop reminding me of how bad I looked.

I got up around noon and took my first walk around his apartment. My leg that was hit gave me fits, but I managed to maneuver rather nicely after some time working out the stiffness of it. I tore off that damned chest binding but left the arm and finger dressings.

I couldn't breathe with that bandage strapped across me like that. Leo bitched a great deal over my taking it off. He eventually relented when I returned to bed and minded him with his other demands. He seemed content with my eating and taking the medications without quarrel. I suppose he decided not to push my tolerance for going against my instincts by ordering I keep it on.

The entire day he played his records and stayed in the room keeping me company. He talked of his youth, Maus, and his travels around the world. I listened with much eagerness reveling in the images he elicited in my dusty "institutionalized" brain.

As he spoke of his visits to the United States of America, I realized how much I had been starving for human companionship. All my years I never had anyone to speak to me other than to command me to get on my knees, serve them, or to threaten me in some way. Leo spoke to me like a friend or brother would. He told me his feelings, dreams, and experiences without acting as if he were better than me.

I was allowed to laugh at his antics without a threat of a thudding, and even ask questions without having to wait to be recognized. I had never really experienced this but a handful of times with Ryker. Even he had been a fleeting moment, almost forgotten by my years of isolation and crushing subjugation.

The day grew long but Leo continued this honest conversation with me. It felt good to be so connected I even allowed Maximillian and Mad Maxx to enter the boy to enjoy it with me and Max. The four of us were spellbound in that bed watching and listening to the greatest person we had ever known.

A light seemed to shine around him causing him to seem angelic and wonderful in every way no matter what folly or stupidity he admitted to us about his history. Leo was just a human being with many imperfections and bad habits. He sat there that day of our convalescence from the third rape letting us get to know the man, not the Dominant.

By the time the sun set, all the shards were hopelessly in love with the schwuler Leo. The one man we had hated only second to Peter. He was not the same man we had met in the Great Hall when he demanded a public blow job after foully examining our back side in front of everyone. None of us would have believed this could have happened had you told us with advanced knowledge.

Not one of the Max boys would deny our love for Leo nor feel ashamed either. It no longer mattered to any of us that he was male not the coveted female of our dreams. He could have been one of Drexel's damned mannequins for all we cared. In our eyes he represented the love, partnership and companionship that had been denied us all our life. Each of us were more than willing to fight to the death just to possess him for our lover for all time.

Leo had done the impossible. He had captured the wild heart of the Priceless Mad Maxx the brutal. Though we all loved him with our very soul, our flesh still was unsure of such a union. Like it or not our sexual attractions were exclusively for the female. Though I had gained the single erection during our minor tryst, I still didn't desire the penetration couple with him.

One can say after years of rape and cruel sodomy situations I had become severely traumatized to anal intercourse and blow jobs. I had to endure them, but that didn't make me ever like a single experience with it, not even the two times I had engaged orally with Leo. I found no joy nor excitement in playing the pincushion nor did having to suck a cock bring me pleasure.

When Leo left to call in our dinner, I wondered if when the time came to sleep with Leo if he would know I didn't find thrill at his penetration. I worried about that a great deal while awaiting his return. I feared losing his affection, but I no longer was able to lie to the man. He had shown me nothing but faith and love. How could I dare to grant him less in return?

He came back smiling with kindness then sat down on the bed next to me, his expression melting immediately. "Christian Axel? What is the matter my love? You always bite your lip when something anxious is on your mind." He ran his hand across my bruised chest frowning.

I chuckled bitterly. "How can you know that? I have only been in your apartment three days, Leo."

Leo snickered "My love, I have watched you since that day in the Great Hall with nothing but interest. I know you as well as anyone can. I tell you something is bothering you, so it bothers me. Out with it. Why are you keeping secrets when I tell you all my own dark deeds?"

I looked at my lap feeling horrid that I was about to be honest with him like this. "Leo, I do love you, there is no doubt but eventually you'll want to couple. I thought you asked me for my truthful passion and not to ever be compliant. I fear my love, when you try to penetrate, I will be nothing but the submissive doing his job. I can close my eyes and forget you're a man if your hand or mouth is on me, but your intercourse, there is no escape for me when that happens. I will only feel despair."

Leo nodded then asked me to look at him. "Christian Axel, listen to me love. I am old enough to be your grandfather, ugly, and worse have already violated you with my finger than forced a blow job against your will. Though neither violation was truthful penetration it was still humiliation for you no doubt. I have always known you to be the straight man. I have no illusions that suddenly I turned you into a raging homosexual. Peter enjoyed the torment your sexual mismatch caused you. Claus, Jonas, and Bladrick, they simply don't care if their penetration bothers you. They view you as their property no matter what they say. Make no mistake, they are as brainwashed as any silver or black in this place. No man can own another. That is bullshit. That said, I told you I need no intercourse with you to love you."

I shook my head feeling my chest grow heavy and it wasn't my broken rib. "I will lose you if I don't allow your couple, Leo. I know you say that doesn't matter but it does. How can I be sure of your heart if I never allow your full expression of it?"

Leo laughed then took my hand looking at the bruises and scars on it. "My love, you have never known anything but rape and fear from sex with a man. I have done enough bad shit in my lifetime. I will not become another cruel experience of exploitation to my love, Christian Axel. I know I must live only with the pleasure of your laugh, smile, and happiness that I now can see in your eyes. I find my satisfaction and pure pleasure that I can boast I gave a little of that to you with so little effort. I say to you I am the luckiest man on Earth to have such a pure love. I will willingly attend to the desires of my flesh without his aid. I swear to you, your love will always be enough to keep me by your side without penetration, a blow job, or even another kiss." He pulled me into a gentle hug.

For some reason his words caused me to break down crying like the frau in his arms. Oh, Meine Liebe, I was just so tired. It had been such a long journey of pain and I still had so far left to go. I think hearing his vow to love me just for me and not what he could take, it broke me down. I had needed to hear these words all my life and until then, never had.

My mother, she doesn't love me, nor does my father. I had no family, no friends, not even a pet to hold. No one cared for Christian Axel, Maximillian, Mad Max, Mad Maxx, Max, nor Christian. Annette had been the closest thing to the real emotion, and she was like Ryker superficial and fleeting. I never got to know either of them well enough to hold them for more than a day. Yet, I had been so starved for affection I had been willing to be abused by Ryker and kill for Annette just to have their love.

I listened through my tears as Leo swore, he would not accept any type of special services from me again unless I initiated them. As I emptied out my trashed-out heart through my eyes, I decided to believe him. If he said he could live without the sex with me, then I would accept that. I didn't know it, but I was finally ready for a true love affair without boundaries. Though it would take more than that promise to overcome years of trauma, incest, and ongoing sexual abuse outside his sanctuary apartment.

That night when Master Leo led me down the hallway, I felt my heart would break having to leave his bed. He turned several times to see me limping behind him. Leo's sadness at saying goodbye for three days was evident. He had even refused to kiss me one last time, saying he would give me only such affection on my return. He never liked the idea of farewells.

Master Jonas came to the door surprised that he didn't have to fetch me from Master Claus's place. Master Leo told him he had just been by there and was doing them this favor, so they didn't have to leave their apartment. He bought the lie and bid me entry. I turned around and watched Master Leo rush away without looking back. I wondered if his hurry was because he was about to drop rain on his cheeks over our parting.

The Vampire took one look at me and ordered me to his bed. I did as told irritated at the reapplication of the chastity device to top off my other miseries. I went to the bed and dropped my bag at the end of it. I started to get into it when Master Jonas came in behind me demanding I strip first.

I nodded and did as I was told keeping my eyes down cast. Master Jonas groaned as he viewed the many cuts, bruises, and other nasty injuries to my flesh from head to toe. I stood there disrobed and trembling from the cold when he came forward wrapping his arms around my waist.

He demanded I look at him. "I have missed you, Christian Axel. It was pure hell without you in my bed. Tell me you missed me too?"

I nodded. "I missed you too Master." I lied like a pro there, but hey he ordered it, ja?

The Vampire smiled his toothy smile. "Good, you are well enough for special services? If so, I request them. I will be gentle, I swear it."

I did my best to hide my disgust and bite back my tears calling out for Maximillian within. "I do whatever you ask of me, Master. Thank you for your mercy."

Maximillian arrived with Mad Maxx and without hesitation entered while Max and I exited. On my way out I shot Maximillian, a look of pity.

"Don't let him get you down. He is only trying to re-mark his territory brother. If he treats you like meat, behave like it." I winked at my wide-eyed brother, then rushed off to get rest before I had to run the wheel once more.

I took the wheel just as Master Jonas pulled the boy into his lustful kissing. I looked over at Mad Maxx shaking my head in pure shock.

"Did you hear Mad Max, brother? What the fuck? Did he just give me advice that was actually useful and not an insult?" I still couldn't be sure I heard that sadist right.

Mad Maxx nodded. "Ja, I heard it too. Maybe there really is something in those pills Leo gives us. Whatever it is, I say don't bother to question it. We have to deal with this couple and with that rib and broken collar bones. Get ready for hell brother."

I snorted. "Am I not always ready for it? Be gentle, he says. Ha! Let me get the broom handle, break his rib, and collar bone, stick it up his ass and say what's the matter? I was gentle, Master."

Mad Maxx and I worked together taking turns at the wheel while the Vampire enforced his intercourse on the boy. I nearly passed out when he roughly forced his cock into our throat during his oral foreplay service forgetting that broken rib hindered our free breath. He did appear to demonstrate a less vigorous thrust but when it comes to being careful, that was the extent of it.

Somehow, I managed to survive it and the man reached his orgasm without lingering too long. I was released to clean up. I happily hit the shower scrubbing off inches of flesh as usual. I told Mad Maxx we needed a wire sponge from the kitchen to make this cleansing ritual go faster. He agreed with me.

When I finished and returned to his bedroom Master Jonas was still in the bed. "Come here, my love. I wish to hold my beloved. You are home at last. It makes this Haus a home when the heart is here."

I limped to the bed and crawled in slowly groaning in much agony from my soreness. My Master grabbed my arm harshly as I slid my legs under the comforter. His eyes were wide with terror.

"Christian Axel, you have done it again. Get up this minute. Gott damn it." He jumped from the bed then grabbed me, pulling me out of it with him scaring the shit out of me.

I cowered to my knees immediately wailing in fear. "Please, mercy Master. I beg of you forgiveness." I had no idea what I had done but this Master was obviously displeased with me for something.

He jerked me off the floor by my upper arm forcing me to stand pointing at my thighs. "You scrubbed all the flesh off your legs and backside Maxximillian. Look at what you do to yourself. Blood is everywhere. This has to stop. I ban you from the showers from this point forward without a Master present to observe. You cannot keep mutilating yourself like this. If I don't stop you, I think you'd take twenty a day and scour yourself till you have no skin anywhere. This obsessive scrubbing will cease. That is a direction." He let go of me.

I whimpered. "I cannot go without a shower Master. Mercy please. The water makes me feel better. It makes the pain go away. I fell back to my knees and came forward into a prostrated position with my forehead on his feet.

Master Jonas shook his head. "Nein, mercy denied. You will not desecrate yourself like this any longer. This disease you have is the cause. You cannot be trusted. I will alert your other Masters to the situation. I have never seen it so bad. I should have stepped in a long time ago. No matter, this stops now. Get up. Come with me to treat these wounds you have made. Gott damn it. You will die of infection at this fucking rate. This schizophrenia is ripping you apart faster than even the Dominant of this Haus could ever hope to. What the hell am I going to do with you?" He waited till I got up and followed him back into the washroom.

I tolerated his cursing and stinging antiseptic sprays. When he was satisfied that I was amply treated he dragged me back to his bedroom by leash. I was ordered to his bed then held tightly in his cuddle. I admit I wept most of the night away. I did my best to stay strong, but I needed that faucet to wash away the horror, the pain, and the madness. Without it, I was condemned to the filthiness of my existence for all time.

The next two days the Vampire insisted I say in the bed. He joined me each night in his coupling and then would haul me by leash to watch me shower. I was beyond miserable. During the day I kept busy trying to learn English and studying my other school subjects, but I needed a shower, areal one, so bad it was driving me crazy.

I managed to puke up all the food and pills he kept trying to feed me. I knew he was messing with my head. I had no idea what his plot was, but I didn't want to be his pawn. I wondered if his keeping me from my showers was part of his plan to drive me to madness. I watched him thinking of how I had been fine until he came into my life. I was sure I didn't have schizophrenia at all. What I did have was a madman that thought himself a Vampire, my husband, and believed farfetched tales that I was a well spring of eternal youth in my veins.

On the day before my return to Leo, Master Jonas left briefly to return a book to Master Claus. He had checked on me before he left. I pretended to be asleep deeply from his pills. He quietly slipped from the apartment. I jumped from the bed and hit the shower turning that water full scalding hot. It burned the flesh and with it went the pain, and nightmares. I gasped with thrill that at last I was clean. I grabbed the rag and soap and tore at my flesh. I watched the blood pour out from the raw skin left behind, feeling immediate relief of my burdens.

I had to hurry so I made sure to take as much of that filthy skin off as possible. I killed the hot water handle with much reservation. I hoped I had gotten it all but without a lot of time, I could have missed some of that foul grime. I jumped out and toweled off wringing out that bloody rag fast as I could. I barely made it back to the bed when I heard the Vampire return.

I let out a sigh of relief thinking that it would be okay now that I washed away the disease once more. The battle to keep it off me was never ending. I need at least five or more showers a day just to keep even with it.

My Master came into the room. I feigned being asleep and he checked the bathroom. I held my breath. He returned with much stealth, then left the room closing it behind him. I opened my eyes and let out my air. He had not caught me. I would need to be careful when around Master Jonas.

I knew he wished me to be ill. Why else was he doing his best to keep me from the cures? I smiled to myself thinking of my beautiful Leo. Only another night to go and then I would be in his arms once more.

That night Master Jonas appeared irritated. His couple was harsher than usual, and he cursed me twice for my "automatic" behavior rather than show any kind of affection. I didn't respond to his complaints. If he wanted to fuck me fine. I only had to lay there and take it. I didn't have to pretend to like it nor love him for it. When he reached his orgasm, he rolled me to my side and laid down next to me staring at my face.

I looked away immediately. I didn't like this bullshit he pulled of pretending his rape was romantic. For him maybe. It was nothing but a literal pain in the ass for me. He reached out and grabbed my ear pulling me to gaze at him.

"Why don't you love me, Christian Axel? I have done everything possible to bring you comfort and joy. I even brought the female into our bedroom. I protect you and make sure you're educated, fed, and clothed. Is there something you desire I do not give you?" He ran his hand down my arm appearing genuinely concerned.

"You give me more than I ask, Master. Thank you for the mercy of it." I said the words that had been beaten into me since my first collaring without inflection or change of tone or feeling of any kind.

He nodded. "Then why? I have given it plenty of time. I cannot be more generous nor love you more. Yet still your heart is cold for your Jonas."

I glared at him. "I am straight, Master. I told you this more than a year ago. I beg your mercy but if you demand special services, I am compelled to grant them or be punished for refusal. You hold my collar and therefore I am your property and prisoner. Forgive me for not loving my jailor. You want a blow job or to violate me with vigor, then I can do what you ask of me. More than that is not fair to ask, nor possible. We went over that at that collaring I seem to recall."

Master Jonas sighed then stroked my cheek. "Then maybe when that collar and leash is gone you can find the expression of joy at my touch, ja? That is all that comes between us?"

I shook my head. "There is an ocean between us Master. When I bust my metal I will run from this place, far away from all the terror and things that are against my nature. I will find my Frau and forget all these nightmares at last. I will hold my children and get lost in their eyes, not yours, not ever."

The Vampire reached out with anger and grabbed my bat collar pulling me close to him by it. "I could kill you if I wanted Christian Axel. Whenever I like I can be cruel, harsh, and brutal. You do not know how violent I can get when things are denied that belong to me. You play a dangerous game thumbing your nose at someone that loves you so deeply."

I scoffed. "Then kill me, I beg of you Master. Do me the fucking mercy. Get another cock warmer for your bed. I will not miss the job. I heard this same bullshit from Master Peter you know. You won't send me to the yard. I have something in my flesh you desire and that is what you love about me. My blood, and my compliance is all you desire for truth. That is all you will ever get from me. I am not insolent when I grant your wishes. I meet them all. This asking for my love is a fantasy that you only make up inside your head. No one can love when the joy of being in another's arms is so fucking lopsided that Christian Axel is turned upside down. You bend over and let me fuck you up the ass with a broken rib, collar bone and looking like the prize fighter, then come speaking to me about fucking love, Master." He backhanded me for yelling like that.

I quickly covered my face and felt the tears of anger falling down my cheeks. He fell back to his pillow letting out his breath slowly. Then sat back up looking at the sheets appearing saddened.

He groaned. "I don't wish to fight with you Christian Axel. I realize this dealing with many Masters and jumping from bed to bed with so many injuries is upsetting. I should have considered this indignity may get to you. I suppose the sexual response I showed was too soon. I couldn't help it. You are so beautiful, I, there is no excuse. Okay, I give you some time to adjust. You are straight. It was too much to hope to capture your heart in such a situation. I admit when I am wrong. I will forget this talk of love and praise you for doing your job well. Then you can at least find peace in it, ja?"

"Would you, Master? I think not. Wait, I apologize that is not proper protocol for one of my low level. Let me correct my insolence this minute. Master, do you need another blow job, or should I get like a dog for your lust? I appreciate the generosity of your telling me how wonderful that feels for you and that I am quite the artist. Forgive me if I must reposition my chastity device so it doesn't chafe my hodensack or get in the way of your pleasures. Your kindness in providing me such peace for the price I pay you is most fair and totally equal. I bow on my knees and thank you for the mercy of not speaking to me for the rest of the night in advance." I rolled over turning my back to him to pretend to sleep like I always did.

The Vampire Master granted me mercy. He laid down and pulled me into his spoon with a growl but said nothing. The rest of the night l laid there and thought of Leo's eyes. When the morning came, I did my best to burn the hours with my books. I kept my eye on the clock trying to will it to go faster. I was packed and ready to make my rotation before nine that night. I managed again to avoid eating the food and taking his pills.

Master Jonas frowned when I walked, my limp had improved to a slight drag, into his living room holding my overnight bag, dressed and ready to go by ten. He looked up from his book with irritation.

"We don't leave for another two hours, Christian Axel. You look like a stick love. I wonder if you are eating enough or if this is some kind of growth spurt? I never seen you look so thin not since the dungeon when the disease onset." He looked me up and down with concern.

I looked at the floor. "I eat plenty Master. Fast metabolism perhaps. I sit over here on the floor and study until time to go to Master Leo's." I dropped my bag and began to take the floor when Master Jonas put down his book.

"Come over here, Christian Axel. You are leaving me for five days and I want to cuddle before I turn over my treasure to the pirates." He patted the couch next to him.

I let out my breath but did as he commanded. He wrapped his arms around my waist then pulled me into his kissing. I endured it but watched the clock hoping the hands would hurry. To my dismay he reached down and undid his breeches. Then he took my hand forcing it to his ready manhood. I did my best to try to get him to take this hand job only.

However, within a few minutes he started pushing my head down demanding I get on my knees to attend his lust with my mouth. I did as commanded without quarrel. He reached his apex without demanding the penetration sex, so I considered myself lucky in that respect. I didn't want to have to hit the shower the minute I got to Leo's, though it was on my agenda, several showers.

I had to endure another half hour of heavy petting and playful biting on my neck from my Master before there was a knock on the door. He sat up not allowing me free of his lap. The Vampire looked at the clock to see it was fifteen minutes before midnight. He growled and cursed under his breath.

"The fucker is early. I apologize to you over this Master in particular, my love. If I had known he would rise instead of the Honorable Hemmel, well I would not have been so quick to agree to the Elders leashing of my man. I am aware what this rat bastard did to you in the name of Peter." He shot a look of hate at the door when the knocking began again.

I shrugged. "I will live Master. Shall I get the door?" I looked to the floor to keep him from seeing my thrill that my Leo had arrived at last.

Master Jonas sighed then ran his hand down my cheek. "You are stronger than I ever gave you credit for my love. A real treasure. If only I could win your love my life would be perfect." The knocking came louder.

"Stop knocking. I am coming, damn it," he yelled, then gently pushed me from his lap.

I got up and rushed for my bag doing all I could to stifle my smile. The Vampire opened the door and Leo stood there appearing irritated.

"Good evening, Count Dracula. Oh – he threw up his hands, flouncing and giggling like the frau queen, ja, a total queen – I always wanted to say the to you Jonas. I am so sorry. I couldn't resist myself. I have come for Mad Maxx. I promise not to suck his blood." He spoke like that fellow in the film that played Bram Stoker's Vampire, you know. Leo giggled at his joke.

I had to cover my mouth from the smile that broke out on my face at his foolishness. I could see Master Jonas was not as humored as the two of us were at this. The Vampire turned to look at me. I dropped my head towards the floor immediately.

"Come to me, my love. The circus has come to town. Look here is the clown. See I can play These little games too, Leo. Now, before you take off with my man you need to be made aware of an issue." He took my leash as I approached, then knelt at his feet.

Leo shot a look of fear at me then gazed back at Master Jonas. "Issue? What has happened?"

Master Jonas shook his head. "Nothing yet. Look you may have noticed that Mad Maxx has a nasty habit of obsessive showering. If you let him, he would take them all day."

Leo nodded. "He is obsessively clean. Noted, I will watch that he doesn't overdo it. Anything else, Jonas?"

The Vampire nodded. "Ja, you also should be aware that after you take your special services rights the boy will scrub all his flesh off from the waist down if you don't stop him. I am sure you noticed this when he was with you last time. Well, that behavior doesn't stop even after he becomes accustomed to your lustful interests."

Leo gasped. "Holy hell, nein, I was unaware of this. I didn't see any of this behavior scrubbing off flesh while at my apartment when he was with me. Is this new?"

Master Jonas squinted his eyes "What do you mean he didn't do this at your place? Mad Maxx has had this problem since the days of Peter's collar. That cocksucker is the one that warned me to watch him about it."

Leo reached out and snatched my leash from Master Jonas's claw. "Well, he likely didn't do this when with me because I have never demanded any special services from him Jonas. I don't go raping little boys like all you do. He had no reason to feel unclean with Leo. I will watch out for the behavior, but I bet I never have to lock my bathroom door. The boy won't be enduring my forced couple not now, not ever." Leo pulled my leash lightly and I started to head out to follow him, but Master Jonas grabbed it holding me still.

"Christian Axel, you answer me truthfully. Is Leo lying to me? Has he never requested your special services?" The Vampire glared at me while I kept my eyes to the floor.

I shook my head. "I dare not reply to that question, Master. It is not my place to answer for what happens in the Haus and bed of another Master. That is forbidden one of my low status. I say this with respect."

Master Jonas smiled. "He won't sleep with you will he, Leo. He refuses you, Ha! I knew it. Christian Axel does love me and only me."

Leo scoffed. "Does he? Well maybe the boy fears you and only you Jonas. Or maybe I respect the face the boy is fucking straight. But who am I to say he doesn't love you. Stranger things than a boy falling in love with the Vampire that rapes and drinks his blood have happened in this Haus. I have seen them with my own eyes. Now, if you don't mind, I am tired. I am an old man, Jonas. I need my beauty sleep. Father time is kicking me in my face and it cracks more every day."

Master Jonas let go of my leash, still smiling widely. He reached down and kissed the top of my head as I walked by then swatted my bottom with glee. This nut really believed that Leo's not forcing himself on me meant I love the Vampire for truth.

I walked out to follow Leo away when in the darkness I saw a tall, dark shadow slip down the stairs. I took a rush toward it. This sudden run of mine startled my Masters.

Leo pulled my leash roughly thinking I was making a rush for the banister. Master Jonas came flying out of his apartment to grab me around the waist restraining my movements.

"What the hell, Christian Axel. See Leo, he is so despondent at leaving his man he is suicidal," yelled out the Vampire.

I struggled doing my best to get closer to the edge to see that fucking rapist. I knew this was him slipping away. He likely come back up here to try to catch me out alone for his lusting again. I was pulled away from my vantage by two incredibly stressed out and misguided Masters.

Leo was gasping and fanning himself when I dropped to a kneel, slipping out of my Master Jonas's grip. I did my best to end their fearing I was trying to leap to my death. Master Jonas stood over me with panic in his face.

"What happened? Why did you do that," He panted out at me.

I dropped my gaze to the floor. "I must have been hallucinating, Master."

Leo let out a yelp. "Oh? What did you see that would cause you to get so upset like that Mad Maxx."

I looked up at Leo with a look of disgust. "The boogie man, Master."

I was allowed to leave with Leo after the two Masters were sure I was no longer hallucinating. I followed my beloved thinking of that cocksucker slinking around at midnight looking to catch me out. Wait, was he stalking Leo? Oh, meine Gott. Could it be jealousy driving these attacks? What if he was planning to kill my Leo.

Chapter 46: The Mission

I followed Leo deep in thought about that shadow lurking on the dark stairs. I trembled thinking that maybe he was planning to kill my Master Leo after he got his fill of tasting my metal. I had to get better faster from his last beating. I couldn't afford to be down and ill. My weakness would give that bastard time to send Leo to join his late brothers Barnim and Drexel.

Leo kept turning around to look at me. I noticed he held my leash tight in his grip. There was a quickness to his steps that made me wonder why the rush. All that was clear is he seemed scared. I swiveled my head all around us trying to see what was upsetting him like that.

When we got into his apartment, he closed the door and locked all his bolts without letting my leash go. I didn't understand this apparent panic in him. I dropped to a kneel feeling fear rising within me. I was thinking Leo knew this monster was following him. Shit. I hoped he had not figured out that the man was violating me as well. This was not good news.

He stood there at his entry watching me kneeling there with my head bowed. I began to shake when several moments passed but he said nothing. I knew it. He no longer loved me, and he was trying to figure out how to tell me. I closed my eyes to brace for the heart break.

I knew it. I should have given in and granted the fucking special services to Leon. I have already been used like a pincushion by several Dominants. What the hell was one more? This is my Gott damned job, like it or not."

At least I was fond of this Master. I could have endured it like all the others. Then he wouldn't be upset with us and ready to toss us out of his affections. Mad Max had been the fool to not call me to attend this coupling. I was sure we had lost the only person that ever seemed to really care about us. I wanted to die so badly.

It was at that moment Mad Max and Max pushed me and my brother Mad Maxx out of the boy. I rolled across the floor grateful for being thrown out for a change. I looked at the sadist and saw he was sweating bullets. His partner Max also looked disturbed. This further frightened the hell out of me. What was happening?

Mad Maxx come over to help me to my feet. "I think Mad Max and Max know you tried to jump from the banister, Maximillian. I think Leo knows it too. We better get the fuck out of here before the shit hits the fan. The tough Mad Max and the soul can handle this mess without our interference. I know like me, after that rough three days with the Vampire and no showers, I am ready to call it quits. We need to be away from the wheel immediately."

I nodded at my brother's wise words. "I wasn't gonna jump brother. I just wanted to get a closer look, is all. I wanted to see if maybe that Barnim is still down there." I looked at the floor.

Mad Maxx frowned then patted my back. "I will only say this one-time brother then we never speak of it again. If that rapist had not come along, you would have jumped and joined the sleeping Barnim below. Tonight, you saw a shadow alright, the reaper came to claim the boy's soul. Maximillian, you are tired. We all are. I think when the Elders Claus and Bladrick come for the flesh, you need to stay asleep. I will handled their gross couple."

I looked up with tears in my eyes that my brother was calling me out on my dark truths. "You would do that for me, brother? I apologize for my weakness. I just want the pain to end. I cannot take this anymore."

The masochist nodded and took my hand. "Ja, we know, Maximillian. You just need rest. Unlike the Masters, Der Hund will grant what it is you require. I handle the London Bridge and even the Vampire until you can shake your sorrows. We are brothers. I got your back." He hugged me and I suddenly felt a little better. Mostly because he didn't seem mad at me for thinking about ending the flesh like I had been for days.

Mad Maxx and I took off for the nothing. It was time for a good long, healing sleep for me, the perfect submissive Maximillian. I am the hardest working of all the shards, and the one with the heaviest burden. I can even wear out when the load gets too heavy. If I fall trying to carry the horrible weight of the boy, then it means doom for us all.

I couldn't believe Maximillian nearly killed us twice. Shit, he had said nothing to any of us about his grieving. I wanted to beat his ass for such stupidity. Max had done his best to calm my anger over this crazy behavior of his, but I admit I was less than over it. Gott damn it. Why the hell did we suffer all we had only to take the flight without wings like that. It made no sense to me, well, okay sure I had been thinking it lately.

Yet, I would never do it. Okay, I will be honest, I was not really sure I wouldn't anymore. something was wrong with all the shards, me included, but of what I couldn't be sure. There seemed to be a rising darkness inside the boy. It was causing a loss of motivation to continue our battle to survive. It was plaguing the flesh to epic proportions the last few times I came to run the wheel. Whatever was going on was more than irritating.

Sometimes it was so bad I thought it may be better to just abandon this sinking ship. Lately, it seemed nothing mattered. All of us had been thinking that even if we did break that metal of ours, we would still be trapped in this nightmare. It was making doing my job harder with each passing hour. That was causing me to wear out too fast. The same damned thing was happening to the shards Maximillian and Mad Maxx.

I noticed that even Max seemed afraid of this whatever coming over us. His eyes told me he was scared to death. I may be a dumb sonofabitch but even I could tell this situation was getting out of hand. I wished to hell I could tell what the issue was. I wanted to fix this shit and move on with the mission more than you can know. Yet, without a clue to the problem, I was helpless to do anything but suffer like my brothers were.

Leo dropped my leash then rubbed his forehead. "Christian Axel, my love, I want to request a favor from you. I pray you love me enough to be honest with your Leo."

I kept my eyes to the floor expecting he would request the special services. I braced myself for it and resolved myself to endure his couple if that what it took to keep his affections.

"I do whatever you ask of me, Master," I said sounding hollow and without emotion in my voice.

He sighed. "I will ignore that programed response, my love. I am asking you as your friend not your Master for this favor. Only you can decide to grant it to me or deny it. You see this door is closed? That means you are not Mad Maxx the Priceless collar anymore. You are Christian Axel the free boy. In this apartment there are no Masters, no subjugation. You and me, are human beings wandering blindly through each day together. I am you equal not your better, my love. Do you understand what I am saying to you?" He stopped to look at me for a response.

I shrugged. "Not really Leo, but I will try if that is what you want." I had no idea how to respond to his insane statement. No Masters, no subjugation, wasn't this his apartment still in the Haus? Did I miss something here?

He groaned and held his head as if it ached. "Fichen mich. This brainwashing shit is beyond my intelligence to fix. Okay my love let me try to say this in a way you can understand, ja? I, your Master Leo, command you to not mind my commands. Mind your own wishes when this door is shut, when you're safe in my apartment, and there is no one else around. It is my desire that you be free in my presence. You do whatever you want, without asking if such a thing will upset me. There are only three exceptions to my order. One, you will take your meds without quarrel, and same goes for eating. I add one more tonight, you will under no circumstances hurt yourself."

I looked up surprised. "Huh? You want me to not mind your commands, Leo?"

He nodded. "Ja. You do whatever you want if it doesn't break my only three directives. Now I ask that favor?"

I was more than a little confused. "As you wish, Leo."

Leo stomped his foot. "Uhg, not as I wish. As you wish, Christian Axel. Oh hell, this will take some work. Never mind, I will help you learn my love. The favor I ask is for you to tell me why you tried to kill yourself back at Jonas's apartment."

I shook my head. "Nein, Leo. I thought I was seeing something in the shadows. That was all," I lied.

Leo frowned. "I see you do not trust me still. That is to be expected. You have no reason to believe in anyone. I know you're not getting those bruises and broken bones from the seizures. I also know you tried to jump from the railing. It is because of the special services with Jonas, isn't it? The scrubbing away of all your flesh, which tells me everything I need to know. Asking you for the reasons you want to kill yourself is like inquiring if it is cold when the snow falls." He looked at the floor appearing saddened.

I shook my head. "Nein. I do not want to kill myself, Leo. I scrub away the diseased flesh only. I am just seeking the cure for my madness, ja?"

Leo looked up with pity in his eyes. "My love you break my heart to hear this truth come from your mouth but at least you are finally being honest with me. Honey, the cure for this madness is not in the shower. Nor is it in the grave. Christian Axel, listen to me when I tell you this, none of it is your fault. You are innocent. These criminals that hold you their prisoner are the ones that should fall from the banister, not the boy I love. You have no reason to feel dirty or ashamed. You do what you must to survive. There is nothing dishonorable in that. You must let go the burden others force upon your shoulders. That is their sin, not yours."

I looked at the floor. "As you say, Leo. Can I please be released to the washroom? I need a shower and to change these foul clothes."

Leo sighed then motioned me to stand when to my shock he knelt to my feet. "You need no permission from me to do as you desire. Please, I beg your mercy that you do not damage yourself by tearing away your beautiful flesh. I would be hurt if you did, but I won't stop you if that is what your conscious tells you to do. That Haus of bone and blood is your own, my love. I can only say whatever cruel thing you do to Christian Axel injures Leo too. If you had jumped tonight, I would have fallen to my death behind you. When you wash away that lovely skin, then I have to see it. It causes me much agony that I helped do this horror to you. I know I deserve to suffer any way you see fit, but I am offering to submit to your punishment for it. I am happy to pay whatever price you ask of me. Take your rightful revenge on the criminal at your feet and stop abusing the boy who never did a thing to earn such brutality."

I stood there; my eyes unable to conceive of his words. "Leo? You don't deserve any punishment or revenge from me. You are kind to me, not cruel."

He scoffed. "Am I, my love? Did you think that when I treated you like an animal in the Great Hall? How about when I threatened to rape you after you suffered in Peter's chains? What about that blow job I threatened you over. You had the choice of sucking my cock or sucking you father's. All in my impulsive pettiness to pay that man back for my perceived slight. Ja, that was real class on my part. Leo really got used by Peter and Cora, didn't he? When the dust settled, I am still an Elder, wealthy, and at the top of the world. How about that boy that we all used? Where is he? He is ready to jump to his death, being raped by old men. and scrubbing off his skin trying to feel clean when such a thing will never be again. I throw myself at your feet and pray your more gracious than I ever have been. I beg your punishment for all I have done against you, but understand when this is done, you must drop it. You can no longer punish Christian Axel for what Leo paid in full." He dropped his face to the floor and began to weep loudly.

I watched the Elder tremble and sob for many minutes. I am the sadist Mad Max. I am without empathy for such displays. Max the soul is not. He listened to the truthful grief of Leo on the floor. I looked to him for his judgement on this bizarre matter. My first instinct was to thud this man to be my twin for all he had indeed done. I never did a fucking thing to him to earn all that he confessed to me. He had earned my right to punish him for his crimes against the boy.

Max smiled at me with a look of peace and calm, then took the wheel from me. "Fair enough, Leo. I am ready to tell you that I do hold anger in my heart for what you did to me. Understand me when I say I believe you are guilty of cruelty, sexual abuse, manipulation, and humiliating me. That said, I am willing to accept an equal return in payment for all you took from me without caring that it would cause me a lifetime of nightmares. Since you grant me the honor of finally receiving what I am owed by you, I grant mercy of the punishment you have without a doubt earned. Your punishment is to help me understand the brutal world that never cared for me. You must suffer this torment with honesty. I am twisted and disabled from this life of abomination you and others forced upon me. Therefore, the payment for your part in my disgrace is that for all your days you must love the monster you helped to create without judgement nor excuse. If you can accept my punishment and pay this price, I demand then rise to endure your punishment and be forgiven." Max held out his hand to the weeping Leo.

Leo raised his face with a stunned expression in his eyes. "I hurt you so much and all you ask in return is the I love you and help you learn about the outside world? My love, how can your mercy be so gentle when you never have known such a thing yourself."

Max chuckled. "Leo, when one never has been loved they are the expert at what it is not. I never received mercy of worth. This I am the most skilled at handing out. However, do realize the punishment and payment I demand is not such a gift. You profess your undying heart, but you don't even know me. do you? The secret is no one does nor can they. Not one person has ever bothered to ask me anything about Christian Axel. They tell me what they want me to say. They project their own desires, lusts, and values onto a boy that is just minding his betters, trying to avoid as much pain as possible. The Masters define me, and I become what they want me to be. Otherwise, I am nothing at all. I will tell you who I am Leo. I am the beast of burden that is looking to escape from his cage without any concept of what is out there beyond the walls. This boy is the God of survival and the notice of living. This empty thing you must love without excuse or disgust. You will reap the bitter crop you sowed Leo. Now, do you accept this as my judgement or were you merely testing my resolve to let your grievous crimes go without fair service return?"

Leo smiled with his tears drying on his cheeks while he took Max's hand. "For a fourteen-year-old, your wisdom is breathtaking my love. I accept the payment and punishment as offered. I give my word that till the grave comes for me, both with be suffered by this worthless clown without regret."

I retook the wheel just as I pulled Leo into a deep passion filled kiss. He wrapped his arms around my waist adoring me with eagerness. I groaned out feeling anxious when he reached down and grabbed my manhood. I was not willing to couple with him nor engage in any kind of special services other than his kissing. I backed away pulling from his embrace and mouth with force. I dropped my face to the floor feeling my ears burning. I hated this feeling that always gripped me before the penetration sex request came from a Master.

Leo gasped at my sudden withdrawal then looked at me with what appeared to be embarrassment. "Oh, meine Gott. Love, I am sorry I didn't mean to frighten you. I was merely checking to see if Jonas put that fucking metal contraption back on. I want to remove it immediately. You deserve your complete freedom in my Haus."

I nodded, still not looking up at him. "Thank you for the mercy, Leo. If you require my services, I must beg you to excuse me for my preparation."

Leo let out a yelp then a loud sigh. "Damn it, Christian Axel. I told you I will never request such a thing from you, and I meant it. You're safe from any lustful touching from Leo unless you ask me to engage you. I think you don't understand this means you will never be commanded to grant me special services. Not even kissing unless you start it and then, if you say nein, I immediately stop. I offer to remove something that is a reminder of your life of subjugation only. If you wish the thing to stay, then so be it. I will further offer you the mercy of sleeping in another room, never having to strip naked for me, and the all privacy any fucking human being deserves. Now, we have this settled. I offer to remove the chastity device. Do you want that thing off, ja or nein?"

I nodded. "Ja, I want it off Leo. Jonas has no right to put it on me. Not one Master has ever tempted me to breach my virginity. It doesn't stop them from taking their lust out on me though. It is stupid and unreasonably cruel."

He snorted. "Fucking right it is. The man is a brutal nut. I go get the hex wrench. This time, with your permission, I will show you how to remove this thing. Next time, you can do it as you wish. I leave the tool where you can have access whenever you like, ja?"

I smiled with my eyes still down. "Ja, thank you Leo."

Leo left for his room. When he returned, he had me drop my breeches then demonstrated the proper way to remove the device without causing pain. He put the thing into a basket on a table with the wrench. He explained this way I had ease of access if ever I felt anxious about having it on or off.

I watched him yawn and pick up his book from the couch as I stood there by the coffee table not sure what to do. "Christian Axel, I need rest. I am going to bed, my love. It has been a long day and I confess my excitement at your coming back to me kept me up last night. I will leave you to decide what you wish to do. I only ask you don't leave the sanctuary of this apartment and remember no injuring yourself. I cannot make you mind either of my requests, but I am asking you nicely and with love in my heart. I will also be honest with you. I have a lock on the door that is not easy to pick, nor will I grant you the key. I know this would only slow you down but not prevent you from escaping if you wanted to do it. I trust you, but not your disease. That is why I had that extra precaution installed this week. You asked me for the truth, so there it is. Good night my love. I see you in the morning." I watched him head to his bedroom not believing my ears that he granted me my freedom to do as I pleased.

I loitered a bit around that coffee table trying to decide what I wanted to do first. Such a thing as free time was novel to me. I looked at the record player and thought maybe of listening to his music. Then I noticed the darkness was rising around my legs. With a startle I headed for his spare washroom grabbing my bag on the way out of the living room. The disease was gaining on me. I had to clean that off.

I stripped with vigor almost mad with terror at the amount of filth that managed to cling to me. Master Jonas's refusal to allow my cure had let it get a tight hold on the flesh. I turned on the hot water and waited with anxiety for the steam to fill the room. I wrung my hands to help keep down the fear within. When the mirror fogged over, I knew the temperature to kill the madness was reached. I jumped into the scalding liquid.

I reached out of the tub for the rag and soap I brought with me. I stood in that stream closing my eyes sighing with relief. In a moment all the scum would be removed from the flesh at last. I opened the rag and took out the piece of sandpaper Maximillian had snagged from Master Jonas's apartment.

My Master had been working on an antique chair he purchased recently. The submissive shard had noticed this most useful rough surface tool sitting next to it. When the Vampire wasn't looking Maximillian had ripped off a hand sized piece of the stuff. I thought if this thing could grind down wood, then flesh shouldn't be a problem.

I applied it to my upper leg and grinded it across the surface of the cherry red skin. I felt the pain as this sandpaper took away the foul tissues. blood rose to the surface and was washed down the drain. I giggled with glee that this shit worked faster than any rag could to remove the disease. I began to sing one of Leo's songs, though I had no idea what the words meant as I was only parroting them, as I scoured away the grime and grossness that clung all over the boy.

When I was sure I had gotten it all I returned my new tool to the rag. I stood in the water allowing it to wash away all the foulness, madness, horror, and sickness of my existence. I nearly felt clean at last. I stepped out of the tub and toweled off ignoring the angry screams from my bloody raw flesh.

When the new skin grew back, a new boy emerged. Maximillian before and now Mad Maxx was renewed each time I killed the defiled flesh that held him. When the Dominants defiled him, I merely would return to destroy that abomination and start the regrowing of a new boy all over again. I had regenerated so many times, I could no longer count how many times I had been reborn this way.

I put on my clothing then applied enough makeup to hide the nasty bruises and cuts from that raping shit. For the moment I was feeling better. My pain was a nuisance, but I was used to always hurting somewhere, you know. I choose to overlook this inconvenience of the broken ribs, collar bone, finger, and all the other injuries to the flesh.

I was beyond thrilled that my sandpaper had managed to do the job so well. Normally, this task took up to five baths a day. I only needed this one to do it with that handy aid. I giggled when I realized I didn't need my razor anymore either.

The Masters ordered me to remain without the marks of manhood such as hairy legs and other more personal places. That caused me the indignity of constant shaving to keep them from punishing me for appearing unattractive for their nasty penetration sex.

Thanks to that sandpaper scrubbing, I noticed my lower half was without a stitch of hair anymore since I took all that nasty flesh off. I wondered if maybe I could destroy enough layers of tissues to keep it from ever growing back. That was something I planned to find out. The idea of never having to deal with the burden of shaving again made me smile at the Mad Maxx staring at me from the mirror. He smiled back with approval at my plan.

It was then I heard the small tapping in the walls. I felt my heart speed up when I realized this was a morse code. This was Der Hund sending me messages from wherever he goes when away. I was excited that he had obtained some information on that rapist.

I cleaned up my mess fast as I could. I did my best to recall all the codes Der Hund was sending me through his taps. I rushed back to the living room and grabbed my pencil and paper. I feverishly worked through the early morning trying to break the codes that surely held the clues to my next target. I was more than ready to get back on with the mission.

When I covered a piece of paper, I wadded it up then put it into my bag. I would take it and drop it into the vent of the room with a grey door the moment I got back home to Master Jonas. That is where I hid all my treasures.

I had the spent syringe that killed Master Stefan, all my private messages from Der Hund, the pills I cheeked from the Masters, broken glass that looked pretty, and even a handful of grass I grabbed the last time I was let outside.

Whenever I got the chance, which was almost never, I would pull these things out and appreciate my wealth. I even had a wrapper and the bubble gum that Ryker gave me one time. I didn't chew this; it was too precious to waste in my nasty mouth. I kept it safe with all my other things to enjoy for all time.

I was laying on my stomach on the floor writing down the codes when Leo came into the room that morning. He stood there watching me a moment. He knelt next to me appearing concerned. I was deeply involved with this complicated process. Der Hund just never told me anything, damn him. Always with the fucking codes and taps. Such bullshit. Not like we were undercover agents or some dumb shit. I didn't even see Leo next to me for several minutes.

"Christian Axel, what is this that you write, my love?" Leo's voice rang in my ears sending me into a screaming cower reflexively.

"Mercy Master. I beg of you. I apologize. It won't happen again," I yelled out as I dropped my pencil covering my head like a pussy.

Leo jumped backward covering his chest as if having a heart attack. "Shit, I am sorry to scare you, my love. I thought you saw me sitting here. It is okay, you done nothing to cause this fear from Leo."

I panted still trying to calm myself from the fright. "Oh, I uhm, apologize but I didn't see you, Leo. Can I serve you somehow?" I pulled my trembling hands away from my head and sat up feeling quite sheepish at my overblown startle.

He came forward reaching toward my face. I closed my eyes and braced for his backhand, dropping my chin slightly. I was shocked when the harsh blow didn't come. Leo stroked my cheek lightly with a tenderness to it.

I opened my eyes to see him looking at me with sadness in his expression. "Oh, this is awful. My love, you are a bundle of nerves. It kills me that your Dominants still feel the need to beat you. I have seen your service. It is polite, graceful, and perfect. What the hell is wrong with them? You are so full of the terror over this cruel behavior I worry you will cower at any noise."

I shrugged. "I am not such a good submissive. I am insolent often, Leo. I know better but sometimes I say or do things that get me thudded."

He nodded. "Well, standing up for yourself shouldn't be considered insolent. Fuck, I swear I near had the heart attack when you jumped like that. You have the reflexes of the cat. Come now, tell me what is this you are working on? I never seen anything like it." He pointed at my coding.

I looked at the floor feeling nervous. "It is nothing Leo."

Leo frowned. "Nothing? Looks like something of importance to you. You can talk to me, Christian Axel. You told me no one ever knows you nor even asks. Well, Leo wants to learn all there is about you. So, tell me what this is you work on?"

I picked up my pencil, still refusing to look up at him. "Uhm, this is the messages that Der Golden Hund sends me. He wants to tell me something, but that boy is complex. I must translate every fucking thing he says."

He nodded. "Ah, I see. Der Golden Hund is a friend of yours?"

I giggled at that poor guess. "Nein, he is the Master of us all."

Leo gasped. "You mean that is what you call Gott?"

I looked at him in irritation. "There is no Gott Leo. Der Hund is the Master of me, all the Max boys and of Christian too. He decides what we do for the mission."

Leo nodded. "Okay ja, I think I am understanding this now. Can I ask what the mission goal is?"

I smiled with happiness that he wanted to know. "The mission is to break the metal. Then we kill Vilber and Olaf to bury outside the Haus doors. We will go to college and become a doctor. After that we find our frau, submit her, and train her to bust her collar. As the Priceless pair we will make children. I hope to convince her to help me burn this fucking place to the ground. This time when the flames do their job, we won't hang around like we did with Gerard. We will run away with her from the police and raise our little ones in the green fields. Maybe we will get a hound and some baby sheep. I don't know if any of us will know anything about caring for the animals though, so that part we are still looking into."

Leo wiped his forehead; he was sweating for some reason. "Oh, that does sound like quite the mission, my love. Uhm, do you think you could eat something for me and take your medication? I am going to order breakfast in this morning. While you and I have the meal, I want to know more about Der Hund and the Max boys, and this Christian. I mean if you would do me the honor of talking with me about them?"

I nodded with joy that someone actually wanted to know about the shards and the mission. No one had ever really asked but the Vampire. Master Jonas was trying to use us in his plots though. We knew that was the only reason he cared to know.

Master Jonas called us masks and said he wanted to enforce his foul lust on each one of the shards. Leo, he had given us the service of telling us about him. If was only fair to provide it back since he requested, it. It seemed strange to us that he didn't seem to have any shards, but we thought maybe he did. He was just not comfortable telling anyone about them.

Leo called in the food and demanded I eat and take his pills. I didn't bother to fight him on either thing since he had been so generous in granting me free time. Then he sat down on his couch after putting on a soothing record. He looked at me still laying on the floor working on my codes from Der Hund.

"Christian Axel, can you come here and sit with me. I want to speak with you about these others you told me about. Let me get to know you and later today I have something special to reward you for it." He patted the couch seat next to him.

I nodded, then put up my things carefully. I sat down next to him and did my best to tell him about the shards and Der Hund. He asked a lot of questions while I told him about my life in the Haus, and before it too. Leo seemed to be captivated by my tale, never moving from his spot to even answer a call of nature once. I was honest with him. I left nothing out, not even the grievous murders Christian had done, or the one Maximillian helped do with that dreadful Barnim.

I knew if Leo wanted to have me put to the yard, he already had enough on me just knowing my truthful identity. If anyone on the Council found out I was really the high born and Priceless son of Agnette and Peter that aided them in raising Cora, I was finished.

I had trusted him with that devastating secret, well blurted it like the dumbass actually. Telling him the truth of my dark dealings in the Haus would not cause worse. after all, even the Guard can only kill you once.

When I finished my story, he sat there with his eyes wide. I looked at the floor wondering if I should be settling my affairs. I began to think telling him all my truths was not such a promising idea. Why I had done it, well obviously I was suicidal. Only someone looking to ride with the reaper would have done such a crazy thing.

Leo took a deep breath. "Well, I see that Vilber and Olaf have their death coming. I never liked that Barnim, Drexel nor fucking Felix. As for Stefan, he is the bastard too, but my love no one has reported him dead yet. You sure about that one?"

I nodded with a smile. "Ja, no one has bothered to check on him yet. You'll see. Once the smell gets so bad, someone will have to come calling. I am the murderous beast, Leo. See I told you loving me would not be easy."

He sighed. "I say good riddance to that collar killer Stefan then. That Milo and Geraldine, oh and Ben, my love, you had no choice in any of these things you have done. I had to kill Helga myself. I am no saint, this you know. You are not a murderer. You are the warrior that survives the battles of this corrupt Haus."

I shrugged. "Then you won't tell on me for the things I have done?"

Leo reached out and took my hand into his with a bitter smile. "I keep my love safe for all time and my mouth shut. You do the same for your Leo. This I swear to you. You did what you were forced to do. As for the shards and Der Hund, amazing adaptation to a nightmare existence darling. I am in awe of your strength to tolerate all you have been forced to endure. I love you even more now that I can boast knowing the real you, all of the you there are. I really am the lucky one. I have not one but six Christian Axels to adore." He chuckled at that.

I narrowed my eyes. "There is only one, Leo?"

He nodded while looking down at my hand. "Ma love, I told you I am blessed. I never knew the terror you have. My mind didn't have to split to survive. I don't have the disease of schizophrenia either. My poor Maus is the one that paid that bill for being born into our family. She was like you are now, my best friend. Maus told me she had many of her like you do. I overlooked the importance of that back then. I no longer am the fool to not see the seriousness of these other Maxes and Christian. I would ask you to do me one more kindness though. Can you speak to the others for me? If I am to adore all of you correctly, I need to know who is at the wheel."

I smiled at that. "That can be arranged, Leo. However, I will be the one at the wheel with you most of the time."

Leo gasped. "Why is that? You are Mad Max the sadist, correct? Does Der Hund think he needs to be protected from me?"

I laughed hard. "Nein, I am here because I am the one that fell in love with you first. Serving you is my pleasure as my reward for it."

Leo chuckled at my laughter. "Wait, you are telling me the sadist of Christian Axel fell in love with the worthless Leo? How is that even possible? I thought you have no empathy?"

I frowned suddenly and glared at Max with his goofy smile. "Max the soul is attached to me. He says I gave him back to Der Hund. He also tells me I am the only one of us that can truly love you the way you love us because of it.

My lack of empathy only means you cannot break my heart if you betray me. I really hate this guy Max following me around but fuck. You should see him fight. I won't cross him that is for damned sure."

Leo nodded. "Ah, I understand completely now. Well, you can tell Der Hund I won't be pissing off Max either. I also won't be attempting to break Mad Max's heart. All of you are welcome in my home. Now, you were kind enough to grant me the pleasure of knowing you for truth. If you are ready, then I am ready to give you that reward I promised." He smiled while looking at his clock.

I shrugged. "As you wish Leo. I go wherever you tell me to go."

He shook his head. "I should correct that statement, but I know that telling you to stop it won't work. Instead of pulling you from this constant kneeling, I am going to teach you how to stand on your own. Come with me love. I apologize but I must take your leash, and you must behave like my submissive for a bit. We are leaving this sanctuary. Outside this door to everyone else I am your Master Leo. Inside this house I am your equal, Leo. Try to never forget no matter where we go, I am your lover Leo always." He told me to get the sunglasses he loaned me and put them on.

When I was ready as he requested, he took my leash and led me out of his apartment. We kept going down the stairs, hallway and right out the front door. I was nervous as I always was back then when in the Haus yard. Leo told me to stay close to him on my leash. He walked me around the back of that place traveling to the distant barns on the grounds.

I could smell the farm animals long before we got to the stables. I realized this was the place I had tried to get to back when Ben fooled me into wearing the dress. I was surprised Leo took me to this place until we went inside.

There by the front of the barn door was a small pen. Five little lambs called out with excitement upon seeing us. I could barely contain myself when Leo opened the door of their cage letting go of my leash. He told me I could go inside and pet them all I liked. I didn't have to be told twice. I was eager to experience what Annette had only given me in her vision of her perfect dream world.

I sat in the hay and petted the lambs. They all wanted my attention. I laughed till I nearly busted in half watching them push each other out of the way. Each was trying to get more than their fair share of my touching. I hugged and adored each of them marveling at their innocent faces and the soft feel of their coats.

Leo stood outside the pen watching me with a large smile on his face. I would look at him whenever another of the babies crawled into my lap wanting his turn for the hugging. I decided Annette was right. I was getting lambs when my Frau and I got our own Haus. They are wonderful animals.

I was sad to leave my new friends when Leo told me it was time to go. I kissed all of them goodbye, then returned to him. He took up my leash and led me out of the stables. I almost ran into him a couple times looking behind me at the lambs that were crying for my return.

He chuckled when we were far enough from the barns that I could no longer hear my friends calling my name anymore. "Christian Axel, settle down my love. You can come see the lambs whenever you like when with me. All you need to do is tell me you want to go, and I will take you."

I smiled with glee. "For truth Master? I want to go back there now. I could stay in that cage with them and never leave."

Leo turned around frowning. "That is sad my love. You make me want to cry with the horror I see in you. You are so starved for love and kindness that you would sit in a stinky barn holding baby lambs rather than a comfortable apartment with your Dominant oppressors. Ah. I cannot face that for the moment. Come with me, I have something else for you to see. You can see the lambs later."

I dropped my head nodding. "As you wish Master." I couldn't hide my disappointment that I wasn't getting to return to the lambs.

He came to me and stroked my cheek. "Don't be upset. I want you to have lots of things of interest to desire. The lambs are only one. There are more, but I do not know what you like if we don't try them all, ja? We go to view another you might find pleasure in. Now come Christian Axel and please be patient with me. This is a learning experience for us both." He took off and I followed, trying hard to forget the lambs so that I could please my Master Leo.

We walked upon a stretch of concrete in the middle of the gardens. On this was a large basketball court. There were about ten black collar males of my age playing this game together. I felt nervous as I followed Leo that walked right up to the edge of that ball game they played.

At first the black collars were so busy with their game they didn't notice Leo and me. Then one of them asked for a moment of rest. He was a tall brunette boy with a crooked nose. This boy pointed us out to his brothers. They all turned to stare at us with curiosity.

Then a boy that was short and heavy with red hair yelled out, "Holy shit, that is Mad Maxx the Priceless collar boys."

Another boy yelled, "Bullshit. No way he would be on the grounds stupid. The Elders would crap a brick if that creepy fucker got out of the Haus."

Then the tall boy said, "The Priceless isn't getting out of the Elders' beds much less the Haus boys. He is far too busy on his knees to be walking. If he had a moment to suck in any air instead of cocks, then he wouldn't be wasting it out here with us. He'd be guzzling the mouthwash."

They all began to laugh, and I dropped my gaze to the ground feeling my ears burning with embarrassment. Truth is, I never was any good at making friends with other people. I was particularly poorly skilled at doing it with those my own age. My Master turned and saw me trying to hide behind him.

He frowned then whispered, "If you want, I can order them to let you play this game with you. You need to have some interaction with other boys my love. This isolation has warped your mind and stifled your social skills."

I came closer to him. "Nein Master. I would rather hold the lambs. Thank you for the mercy."

He nodded. "Okay, maybe this is not your scene. We try other possibilities, ja?"

I gasped then moved closer. watching with fear as the black collar youths started walking our way. I was not in the mood for an ass whipping so close to my last one by the rapist.

The tall boy stopped just as he got within good viewing distance. "I will be damned. It is the Priceless. Oh, meine Gott. Leo, sir, me, and the boys apologize for our rudeness. We didn't realize this was you." He put up his hand to motion the other boys to halt their march to get a closer look.

Leo scoffed. "Ja it is me and the Priceless Mad Maxx. I had brought him here boasting of the good manners and skills of Edgar and his black collar boys. I am most offended to find my brag in error. You should all be ashamed of yourselves. I guess all of you forgot that to offend a Priceless carries a punishment of the dungeon or even death."

The boys all gasped in terror realizing Leo was correct. They all fell to their knees wailing in terror begging for Leo's forgiveness over their foul, but truthful, statements. Leo turned to look at me then back at the freaked-out boys.

"All of you shut up. I give the right to decide your fate up to the one you have dishonored. You speak without knowing. Now you can taste the wrath of one that can offer you mercy or crush you with that mouth you all just made foul and incorrect accusations against." Leo pulled my leash lightly and motioned for me to address the terrified teens.

I stood there looking at them. They all dropped their eyes to the court and trembled. I thought of all the times I myself had been punished for just being the boy I was. My Max nodded at my decision.

I growled out. "I grant the mercy of my silence. I hear nothing here of interest from these black collars. In fact, the magpies in the fields speak with more value to my ears. I hope the worthless boys that have been so kindly treated today remember to grant a service return. I would expect their own tongues to be as still as mine about gossip that has no validity. I am ready to go, Master. This area is without a desirable view." I returned to my place of high protocol behind Leo.

Leo shot me a wicked smile, then addressed the relieved black collars still on their knees "Well, looks like the brutal Mad Maxx is more forgiving than Leo can ever be. You are lucky I granted him this right. Had I chosen, you'd all be playing a game of basketball with Gott tonight. Come Mad Maxx, I agree with you love. There is nothing worth looking at around here. Just trash far as the eye can see." He walked across the court, and I followed while the boys kept their heads down panting in fear that they had avoided the Guard.

Leo and I giggled a great deal once we got out of earshot of the boys. He puffed up his chest and shouted out the he would have them all in Gott's ball court by midnight in a deep masculine voice. I pretended to be afraid of his threats and that made him laugh even harder.

We were nearly out of breath from our joking as he approached our next destination. To my thrill a set of swings, a slide, and a merry-go-round appeared in the middle of a well-manicured garden. He let go of my leash and told me to enjoy These pleasures that had been denied me in my childhood. I rushed to the swings forgetting about the snipers. He had to run after me to keep the bullets from ringing out.

After catching up he gave me the gentle reminder that I was being watched. He then let me get in one of the swings and pushed me. He spoke to me of his own days on the playgrounds and told me his favorite sport was the football, oh wait, you call this soccer. *This game you know Meine Liebe. Nein? Well, that is sad. We fix that another day, ja?*

Anyway, I enjoyed the afternoon with the wind in my face swinging like a kid. Leo eventually talked me into trying the merry-go-round. I got on that thing, and he pushed it to amazing speeds. I laid down on it and watched the sky spinning. I kind of found this unsettling. I didn't need that equipment to see this very thing. I needed only to miss my medications.

Leo had stopped pushing to catch his breath. I laid their thinking of the lambs. I wondered if Leo would be angry if I asked to see them before we went back into the Haus. It was then I saw the face of a man I remembered from my life under Peter's collar. He was standing next to the merry-go-round smiling with the same grin I saw as Xavier tortured me for his, and others, viewing pleasure. I sat up immediately in terror. There was Julius watching me spin.

Leo come over to block the man from getting any closer as the machine began to slow. I kept my eyes on this horrible Dominant craning my neck to near pain. I heard him begin to chuckle the way he had that night so long ago.

"Leo, I see you have the Priceless with you. Ah, he is looking a bit roughed up brother. I thought Xavier died. Hell, I know he did. I saw it with my own eyes. This boy's couple was too much for that old poser. Get him off that children's equipment. I want to get a closer look at him. He has grown. Damn he will be a big one, ja?" Julius put his hand into his front pocket.

Leo scoffed. "Julius you can look all you like from afar. Mad Maxx is the property of the Elders. You keep your hands off him." He crossed his arms as the merry-go-round finally started to slow to a stop.

Julius pulled out a small book. "Get off it, Leo. Everything in this Haus has a price. How much for a ride on the Priceless's merry-go-round?" He chuckled at his stupid sexual innuendo.

Leo let out a yelp and snapped his fingers with indignation (ja need I say it, Queen Leo). "How dare you. This collar's rights are not for sale for any price. Julius you pig. Get out of my sight before I lose my temper."

Julius grinned at Leo, "Oh well, no one can say I didn't at least attempt to get my thrill the legal way. Tell you what my little Priceless collar is. You just meet me in the storage closet, or behind the backstairs and I will show you what a real man can do to you." He winked at me sending shivers down my spine. The rapist was Julius. How else could he have known what I told no one.

Leo pointed toward the Haus. "Beat it, Julius. You're scaring my Maxx."

Julius nodded, still smiling. "I do intend to do both. You better watch your back Leo. I believe I owe you a little something for that shit you pulled on me. Buh-bye, Mad Maxx. See you soon. You boys have a nice day." He turned and took off width speed toward the Haus.

I jumped off the machine looking at Leo in total terror. "Master? This man believes you have done something to him?"

Leo snorted. "I did nothing to the creep. That fucking girlfriend bitch of his Beatrice, she was abusing the female silvers. The horrid thing was buying the simple metals cheaply then selling them out to the Guard as playthings for more. When they finished with the little girls, they were fully soiled and then some. She would then send them out on circuit till they were no more. That shit is forbidden. That whore was nothing but a predatory psychopath like Julius there. The Council looked into my complaint and found her guilty, fair and square. She was exiled from the Haus, which was too fucking good for the monster. Anyway, old Julius met her in his Haus in Belgium. You know how lovers can be. He has taken up Beatrice's cause. This is not the first threat he has given me. You stay away from that bastard, my love. This fight is between him and I. You need not burden yourself with it." He took my leash and began to lead me back to the Haus.

I shook even though it was not cold "Master, that is not correct. Your enemies are Mad Maxx's."

He chuckled. "Ah, you are too good to be true, my love. Nein, you let me handle my own battles. I believe you saw my work. I can make Julius and the devil sorry he ever met Leo. Ask Helga."

I nodded then sent a message to Der Hund that Julius was the target, not Grisham. Julius was the rapist. He knew the locations of the rapes, indicated he would see me for a beating soon, and had a deep hatred for Leo. Julius was on the voting Council as well. He matched the description and without a doubt had been trying to gain a taste of my metal for almost as long as I had known what special services were.

I followed Leo back into the Haus feeling anxious that soon I would need to have a plan to murder this man. Allowing him to live much longer risked both my safety, and based on his statements, those of my beloved Leo. I couldn't bear the thought of losing him. I had never been treated so well with such love, nor more fairly in my life than I had been by this Master.

When we got back to the apartment Master Jonas was lingering outside Leo's door. I saw him and dropped my gaze worried the Vampire had come to start more trouble with the Elder. Leo groaned, then shot me a look of caution.

"I don't know what Jonas is up to, but you be silent my love. Let me handle this. You are on my clock. He has no right to bust into my time with you. Stay calm no matter what he says." He turned around with a big fake smile on his face.

Master Jonas frowned. "This is not a social call Leo. I have a problem and I need you to help me with it."

Leo walked past the Vampire pretending not to be interested. "Oh? Depends on what you need Jonas. I am a busy man. Please state your business, I have things to do."

The Vampire looked at me with a look of desire. "I have been notified by family there has been a death. I must travel a distance to attend the funeral and the reading of the will. There is quite a large estate to settle. I will be gone two weeks. I spoke to Claus and asked him to give you my time with Christian Axel until I return." I looked up at him with a startle.

Leo stopped trying to unlock his door, appearing as stunned as me. "Huh? You ask them to grant me your days in the cycle? Why would you do that? I would think the others would want to share them equally with me and you desire that as well. However, I do thank you for honoring me by showing trust I can care for Maxx. Oh, and my sympathies for your loss."

Master Jonas chuckled. "I didn't know the member that died but I am named as the administrator of his estate. He likely named me because I was unknown to him. My family is a pack of buzzards. I asked to have you hold my man while I am away because he refuses to grant you special services. I know he hates you. I apologize Christian Axel for putting you in the apartment of this man that did you no favors, but it is better than running from Claus's wanton lust, ja?"

I nodded then dropped my gaze trying to stifle my looks of joy that I would get to stay with Leo and avoid the Vampires lusty grip for two weeks. "Ja, I thank you for the mercy of your wisdom Master." He reached out and ruffled my hair with a toothy grin.

Leo put his hands on his hips dramatically acting offended, he was full of shit of course. "Well, I will accept this burden, but I promise you I will expect you to pay back the favor. Nothing worse than having a beautiful boy laying around naked all day to tempt me. You know I won't rape no one, but I admit my hand does tire of thudding him over his constant refusals. I gave up on him. Damn hardheaded little bastard. No matter how much I threaten, he won't let me near him. Fuck it I say. There are other little fish in the sea." He snapped his finger wiggling his bottom with much vigor.

Master Jonas rolled his eyes. "Ja, ja Leo. Do you mind if I say goodbye to my man? I ask this politely though it is my right." He reached out grabbing my leash pulling me toward him.

Leo shot me a look and I winced but nodded that I was okay with this. "Uhm, okay but make this quick. I need to give him medication and feed him."

Master Jonas laughed. "If you give me a minute, I will fill his stomach for you, right Christian Axel."

I groaned. "As you wish Master."

Leo gasped. "Oh, hell no. You kiss him all you like but none of the heavy shit on my time Jonas. If I cannot have this boy, then you sure as hell cannot."

The Vampire growled. "You always were a bitch, Leo. Come here and kiss your man goodbye, my love. I will just have to wait to have you the way I want when I return.

Maybe a few weeks without any affection will change your attitude towards your Master Jonas, ja?"

I shrugged "Whatever you say, Master. I do what you ask and thank you for the mercy of it." He pulled me close and kissed my mouth with much vigor and pawed without care. Thankfully, he didn't discover my chastity device was missing since he wasn't groping me there. But he did take my hand, forcing me to touch him. I can say this was a case of "whew" and "yuck" at the same time.

When he finally had his fill of this humiliation, he let me go and left as if in a hurry. I shot a look at Leo that had done his best to look away from that vulgar display of the Vampire's wantonness. He didn't say a thing. He opened his door then led me inside. I flinched when he slammed it shut. Then threw his keys across the room. I drop to a kneel, assuming he was angered by watching me attend the lustful molestations of another.

I kept my eyes down while he flounced and cursed Master Jonas. Then he noticed I was trembling and got silent. He looked at me with an expression of apology.

"My love, you need not be afraid nor kneel in this Haus. I am not angry with you. I am angry at me for being helpless to stop the man and others like him from the hurt they cause you." He dropped his gaze to the floor.

I shook my head. "You need not worry, Leo. I thought of the lambs when he did that foulness. I thank you for that mercy. I am used to this unwanted touching business. May I please use your washroom? I wish to brush my teeth."

He sighed. "Honey, I told you there is no need to ask me permission when this door is closed. By all means, brush your teeth. I would want too as well if that brute did that shit to me. I will order our dinner, while you do your hygiene. We can open that new record of mine and listen to it over our meal if you wish?"

I nodded. "If this pleases you, Leo. I will hurry." I took off leaving the still fuming Leo by his front door.

I grabbed my bag and took out the rag and soap. I started the hot water tap and stripped fast as possible. Luckily, that sandpaper assured this cleaning would be rapid. I waited for the steam to roll out then got into the scalding water.

Th pain stopped aching in my chest as I was overcome by the blinding white agony of the heat on the flesh. I breathed out in relief, feeling calmer almost instantly. I could feel it burning away all the filth of my disease, at least for the moment. I opened the rag and took out my new tool.

I had started the sanding when the door flew open. To my surprise Leo came rushing inside the room. I didn't have time to hide the paper when he pulled back the curtain. He let out a scream of terror.

"Oh, meine Gott, Christian Axel. What have you done to yourself? Nein, stop this my love. You have nothing to be ashamed of. You are not unclean from the filthy actions that others force on you. Please give me the sandpaper." He held out his hand unable to tear his eyes off my skinned legs and hind end.

I dropped my head and watched the fresh blood rolling down the drain/ "Please Leo, leave me be. This is the only way I can manage this dishonor. Don't take my cure. The filth will corrupt me. I need the water. It makes the pain go away. I thought you said you understand." I held tight to the sanding tool terrified Leo would hurt me like everyone else did.

Tears broke out on Leo's face. "My love, look at me. Give me the sandpaper and I will not interfere with your way of dealing with this horror you must endure. I cannot allow you to rip yourself apart like this though. I will let you stay in this shower for all the two weeks if you want if you give this thing to me." He stuck his open palm closer to me.

I looked at the paper, then back at his hand. "I need this, Leo. I cannot give it to you."

He nodded then smiled bitterly. "I will take you to see the lambs all morning tomorrow if you hand it over. You can stay in this tub all night till we go. I won't say a word. Please now love, give it to me I beg of you."

I leaned my head against the wall and gave the sandpaper to Leo. He took it and appeared most relieved by it. I understood that. It sure gave me peace. I would have to steal some more from Master Jonas when he came back. Until then, long, hot showers and constant shaving was back on my schedule. Leo left the room reminding me I could see the lambs in the morning. He told me he would bring my meal into the shower if I didn't wish to come eat with him in the living room. Then he closed the curtain and left the bathroom, closing the door behind him.

I sat down in the tub letting the hot water run over my head. I didn't know how to make this all stop. The madness, the confusion, the constant struggle with my disease. It seemed hopeless. I looked at Max. He shook his head indicating he didn't have the answer to any of this either. I was just so tired of it all.

I heard the song "Nights in White Satin" playing through the wall. I turned off the facet so I could listen better. I laid in that ceramic bath for the whole thing wishing I had a way to make that sound real and this life I lived the recording. When the song finished, Leo played it again.

I crawled out of the tub and dried off. I redressed quickly and brushed my teeth vigorously. I packed up my hygiene things into the bag and cleaned up my mess. Leo put the needle on repeat. That beautiful melody rang out for a third time. I knew this was Leo calling me to join him in his living room. He had been so kind and thoughtful to me; how could I tell him nein?

I came out of the hallway into his living room to find him sitting on his couch with his hands over his eyes. I could tell he was deeply upset. This bothered me a great deal. I didn't know how to fix it. I stood there like the dumbass, unsure how to proceed when he spoke.

"My love. Come join me on the couch if you want. I do wish you to eat your dinner and take the medications for me." He didn't take his hand from his face.

I looked at the floor. "Are you angry with me, Leo?" I felt I may cry thinking it was all my fault he was grieving this way.

He looked up at me with an expression of sadness. "Nein, Christian Axel. Not at all my love. In fact, you make me very happy. I am the luckiest man on the planet to have such a wonderful companion to share my dinner and this Haus with. Come sit and sup with me. We can listen to another tune if you like, or I will turn it off all together and we can just talk to each other like people do, ja?"

I nodded. "Whatever pleases you, pleases me, Leo. I wanted to ask if you were serious about the lambs?"

He smiled. "You bet I was. We will go see them the minute we wake up. I bet we can even get the black collar that tends them to let you feed dem."

That made me smile with much joy. "I would like that a lot."

I sat down and ate the meal as he asked and took the pills he gave me. I listed to his new record by a group called the "Eagles." He translated the lyrics and helped me practice my English. I was still very deficit in the language, but he made a big commotion as if I was learning quickly.

After dinner, I went to the floor and worked on my school studies. He read a book and sighed from time to time over it. He told me it was a romance novel and the scenes made him swoon. I had to chuckle over that crazy business.

At eleven he bid me good night and went to bed without any demands of kissing or petting. I watched him yawning as he went inside his room. I went back to my work but couldn't stop thinking about how much he meant to me. The more I got to know him, the more I couldn't imagine my life without him in it. I cursed not getting to know him better sooner and wondered if I would not have suffered as much had I bothered to look beyond what I thought I saw.

Around one o'clock I felt a little tired. I went to the couch and laid down. I tossed and turned feeling most anxious. I couldn't stop thinking about Leo. I gave up trying to sleep on that uncomfortable thing. I decided to try out an idea to gain a little time of relaxation off the wheel.

I snuck down the hallway and entered Leo's room. I saw he was fast asleep on the right side of his bed. I quietly slipped into it on the left. Then very gently put my arm around his chest. He stirred slightly but didn't wake up. I laid there watching him for some time before my eyelids grew heavy. I was nearly asleep when Leo rolled and noticed my arm wrapped around him. He saw me there.

With a surprised look he called out. "Christian Axel, you didn't have to come sleep with me. There is another bed in the apartment or even the couch."

I nodded. "Ja, I know. I prefer this one."

He smiled as I kissed his lips lightly. "You are welcome in my bed anytime love. Now get some rest. The lambs will be eager to have their breakfast."

I yawned. "I won't let them down." Leo put his arm around me, and I rolled to allow the spooning.

I feel into a deep, peaceful, dreamless sleep for the first time since, hell I never had that kind of slumber before, who am I kidding? It was a good thing too because I was just so fucking tired of it all.

Chapter 47: Leo's Lambs and the Unblemished Friend

I awoke to Leo's alarm the just as the sun rose. I was eager to get back to see my lamb friends. I was gentle in waking the Elder. He laughed at my jumping from the bed to put on my boots excited to leave. He gave me a calm warning that we were going nowhere until I let him get dressed and a cup of coffee.

"You are a young boy, my love. Leo is the old hag. I cannot just hop from my slumber and run off to the stables like you can." He ruffled my hair when I dropped to a knee at his feet.

"I will aid you to dress and get your coffee, Leo. Then we will move faster, ja?" I did my best to not show my irritation that he was wasting time by bothering to yap with me.

He frowned. "I can dress myself love. You may go get me that coffee to make things go quicker but no more kneeling to me in this apartment when it is just you and me. I don't care for it."

I raised an eyebrow at that. "As you say Leo." I got up and rushed off to fix his drink wondering why he was so weird. *It took me all that time to learn to kneel correctly and now this Master says he doesn't even like it? Why the hell did I get so many thuds over it then. I told you I never understood Dominants, not to this day.*

When I returned with Leo's coffee, I found him struggling with his shoelaces. I rushed over dropping to my knees to tie them for him. I noticed his left-hand was stiff and several of the knuckles swollen. This was causing him fits over getting his shoes on.

He saw me looking at that hand. "Ah, I broke this hand years ago in a horseback riding incident, my love. I have terrible arthritis in the damned thing now. The accidents of our youth catch up. One day you will understand." He stroked my cheek and thanked me for my aid and the coffee.

I stared at the floor not rising. "I already understand, Leo. I have much pain even when the cuts and swelling heal already. When I am old, if I survive, then I will be in a wheelchair, ja?" I trembled at the thought of that.

Leo grimaced. "I don't think so. They have new discoveries in medical science every day, my love. When you are the old man, we will live on Mars, or another planet possibly." He shot me a bitter smile.

I shook my head. "I would like to just get to stay on this one, Leo. I am tired of my trips to outer space. I must say, if they ever offer you the chance to go to the stars, tell them nein. It is not a place anyone should wish to ever be."

Leo snickered. "I love your sense of humor, Christian Axel. You say these things like you are serious, but I have already learned to know when you are funning Leo."

I looked up into his laughing face with humor. "Ja, I was kidding, but how did you know this?"

He stood up and gulped from his cup then said, "You always cross the fingers on your right hand. Come now, my little lamb. Your brothers are calling out for you to feed them." I jumped to my feet as he took my leash and led me out of the Haus right to the stables.

Leo opened the door of the pen while I rushed inside to the happy babies. He told me not to move from that cage. He needed to locate the attending black collar. I nodded but wasn't really listening to him. All I cared about was the soft, pretty lambs that came fighting each other to gain my love and kisses.

I cuddled them all, sometimes three at a time. I spoke to them of the playground and those black collar youth that were not fond of me. But I lied and told them I was their buddy. The lambs listened to me with much intensity. I knew they were grateful to hear of the world outside their pen. It is not fun to be in a cage. That I understood very well.

Leo returned with a portly, black collar man that had a big neck and was sweaty. He pointed at me and the lambs saying something. The man nodded then took off with speed. Leo come to the edge of the pen smiling.

"This black collar Rudolf, he attends the lambs. He is going to get their feed and says he is happy to give this chore to you." Leo giggled when I turned to look at him holding three of the lambs in my lap with much clumsiness.

I smiled "I can feed them every they for breakfast?"

He frowned. "Uhm, as long as you are with Leo, ja. You will not always be with me, my love."

I nodded, still smiling. "But I will give it to them three days a week. That will keep them remembering Maxx that loves them."

The black collar showed up with their grain. I took the bucket, and the baby sheep nearly knocked me to my ass fighting to get that stuff. I giggled and watched with awe as their little lips wiggled while that ate. I got down on the ground laying in the hay to get a better look.

I heard Leo ask Rudolf if he would mind allowing me to feed them whenever he called ahead. The man nodded then chucked with humor.

"Sure, Leo not a problem I am happy to let Mad Maxx feed them. One less chore for me, but I have to say I was near done with this chore anyway. These little beauties are gone in two weeks." He wiped his forehead.

I sat up with a startle. "Where do they go? Back to their mothers?" I was sad to think I wouldn't see them anymore in the fields, but it was wrong of me to want them in this cage just for my pleasure.

That made Rudolf laughed out loud with a look of surprise on his face. "Nein. Their mothers, are you kidding?

These lambs were taken from their mothers, Mad Maxx, and will never return to them."

I frowned at that. "Their mothers were not good to them. I suppose I figured that out with them here and not with the others. They are so little though. Won't the bigger sheep in the flock hurt them when they go home?"

Rudolf shot a look of incredulous mirth at Leo then back to me. "Mad Maxx, the big sheep will never get the chance to hurt them. These lambs are for the Dominants lamb chops. In two weeks, they go for slaughter. Is this some kind of joke, Leo? Surely, this Priceless is funning me."

I let out a wail and grabbed the lambs closest to me in terror. "Nein, Master. Don't let them kill these lambs. I will murder the first person that steps in here with the axe. I swear this."

Leo saw that I was getting agitated as he watched me trying to pull the lambs to me and hide them in my coat. "Maxx, calm down this minute. That is a directive." Rudolf backed away from the pen, his humor melting into fear as I looked at him swearing to end his days for hurting my friends.

Leo smacked the side of the cage. "I mean this, Maxx. You stop this insolence or downstairs we go."

I shook my head wildly. "You can take me, Master. I will earn that beating. Rudolph, you come in here and fight me like a man. You're a pussy. You like killing little helpless lambs, but you are afraid of the one that can hit you back." I stood up and took the stance of aggression daring the worthless black collar to engage in the battle with me.

Leo gasped, then took a deep breath. "Listen to me, Maxx. I mean it. You drop this behavior this minute. If you are willing to grant me a favor, any favor I demand, then I will buy these lambs for you. This will mean you must take care of them, or they will not survive. You will be responsible for their good health. Do you agree to this arrangement ja or nein?"

I didn't drop my fists but nodded. "Ja, I take this deal Master. Only if you can swear no one will be allowed to kill them for fucking dinner. Let them eat fucking chicken. It is healthier anyway."

Leo sighed then looked at Rudolf. "I pay any price for the five of them. These lambs are now only for Leo and only you can touch them. If anything happens to them, it will be your neck."

Rudolf nodded. "I understand, Leo. I will take fine care of Leo's lambs. Thank you for the honor of this." He bowed then left to make the arrangements for the title transfer of these babies to the Elder.

Leo looked back at me with sternness. "There, my love. The lambs are safe for all their lives. No one will dare to injure the property of the Elder Leo. You can stop acting the fool this minute."

I dropped my fists and my head. "I cannot lie by apologizing Master. I am not sorry for what I did. I am grateful for your mercy to the lambs though. I demand I be punished for my insolence." I felt the lambs nudging my knees almost like the knew I had saved them from their slaughter.

Leo shook his head then spoke to me with fierceness in his voice "Your punishment is that you will not scrub that flesh away from your legs while in my apartment anymore. You can have one shower a day only. You refuse this punishment I call back Rudolf right this minute and change my mind about the lambs."

I grimaced with fear in my heart but the babies at my feet depended on my compliance with this terrible price I would have to pay for them. "As you wish, Master. Thank you for the mercy."

He frowned. "There is no mercy in my punishment for you, but this must be done. Now I name that favor you agreed to. This is not negotiable either. You will finish with your lambs' breakfast then we will go home instead of staying all morning as planned. I am calling the Haus doctor and having your flesh looked at from your rough treatment with that sandpaper. No arguments."

I groaned. "But Master, I don't need to see this doctor. There is no reason to be afraid. I got all the diseased flesh, I swear this."

He smacked the side of the pen making me flinch. "I think I am hallucinating. I heard you arguing with me. I hope not. I said no arguments. I am not being unfair in this. You are not well Maxx. That flesh is to be looked at and assessed for treatment. I give you five lambs that will depend on you to keep them safe. They are helpless to defend themselves from the butcher. Only you can do that for them. Something happens to you from infection or injury, then they die and get served on the plates of the Dominants. Is that what you want my love? I am not willing to care for these animals. I have honored you by granting your request to take this heavy responsibility. If you don't wish to do it, then throw them away and do as you please in the shower. You can refuse the doctor's treatment; hell, even jump from the banister like a damned fool. You see, my love, you need to learn to make choices for yourself without threat to you personally. Do remember nothing of worth is free Maxx. That is a lesson in this. You can attend to the lives you claim as important and meet my price for it, or you can go free without care. They will pay the price instead of you but that is life, isn't it? I will not judge you either way you go. This is completely up to you. So? Who will pay this bill that is due, you or them?"

I hung my head lower and nodded. "I will pay this bill, Master. I want the lambs to be safe. I will take the responsibility for their good care. I will accept the punishment and price as you stated without further quarrel over it. Thank you for the mercy of your wisdom and patience with this unworthy submissive."

His expression of anger softened. "You are most wordy, my love. I would not do this for anyone else on this Earth. The Haus accountant will rip me off on the price of these beauties, smelling my weak position, However, the smile these animals put on your face and the bounce in your walk is worth anything I must pay for it. You can send Der Hund a message and you listen closely Mad Max, if any of you were thinking of ending the boy be aware I will not attend these lambs. You must live so they can too. That is what responsibility is all about. Tell me, do you understand this?"

I smiled as I petted my children, the lambs on their heads. "Ja, I do Master. I wasn't thinking of ending the flesh, I swear it."

He scoffed. "You are a terrible liar, my love. You stay here and play with your lambs. I go to pay the king's ransom for them. Do not move from this pen. I will be back shortly." I nodded and sat back down to do as he told me while he stormed off to find Rudolf.

I giggled with joy that all these souls were mine for truth. I immediately sexed them and found I had three boys and two girls. I decided they should all have names so they would know when I talk with them and not their brothers or sisters.

The one with the wobbly legs I called her Annette because she fell in my lap often. The other female was strong and bossy. I called her Geraldine. One of the male's was the bully like my lamb Geraldine so I called him Ryker. He thought himself the Prince of them all. He made me laugh often by acting the fool when he was no bigger than any of them.

One of the male's was the prettiest of all them. His baby wool was sleek and softest. I called him Milo. The last male was the friendliest and most patient. I called him Abelard. I hugged and kissed each one when they let me know they appreciated the names I gave them.

I was feeling like the luckiest boy anywhere. I had five friends that loved me no matter how disgusting I was. If they thought poorly of me for being the nothing, then they did me the mercy of never telling me of it. I thought if there had been a heaven for truth, it would have to be full of lambs.

When Leo came back this time, he told me I had to follow him back for my repayment to him. I sighed and gave Abelard, Geraldine, Annette, Ryker, and Milo strict orders to follow Rudolf's directives, or I would hear of it.

I knew my lambs were polite and well behaved, but I didn't want to seem too soft. I knew if they became too spoiled from my permissive parenting then the other sheep would never accept them into their flock.

Leo chucked as he listened to me give my lambs their instructions. I shot him a look of caution. He covered his mouth and let me continue till I was sure they all understood. He took my leash and led me back to the apartment. I listened to him tell me that the accountant had not been too steep with the price after all.

Not when he heard the lambs were a love gift to the Priceless collar from the Elder Leo. He told me that it sickened him how many in the Haus, Dominant or black collar, were willing to bend rules when they thought it would bring them future favor.

I listened to him talk of the way the real world works without interference. It seemed to me outside the walls people were not really much different then within them. Like the Dominants, black collars, then the lowly silvers, there were those with much, some that had a little, and other that had nothing.

At the end of the day, the ones on the top always used the nothings at the bottom. The middle types too often used the ones without, but they often accused the wealthy of doing all the ills of society. The way I saw it, I was not going to be any better off when my metal was busted. I would be raised to be a low born Dominant in the Haus, but in the world, I would still be a nothing.

These thoughts were in my mind when we arrived back at his apartment. I went to the couch and sat quietly while Leo made his call to the doctor. I wanted badly to go to the shower but decided if I was to only get the one, I would have to wait till I was sure that all chances of foul behavior had passed. I decided to take it at eleven to one in the morning every day under Leo's leash. This seemed like the best way to deal with the battle against my filthy disease.

The doctor had me strip from the waist down. He gasped with terror and asked Leo a lot of questions about how my flesh had come to be in this condition. I sat there irritated at the silly concern they both seemed to be demonstrating over my removal of this disgusting flesh.

The doctor treated my wounds with stinging antiseptic then wrapped my bottom half in gaze and bandages.

He told Leo I would need these treatments for another two days at the very least. I was thrilled when he said my flesh was so severely damaged, he thought my hair in many places was unlikely to grow back. *Dat was good news, and for the record, Meine Liebe, he was correct. I stopped and looked at his legs realizing they were not shaved like I had thought them to be. They were scarred so badly their hair couldn't grow there.*

When he finished his job, I was forced to sit quietly as he told Leo I was not to get any of these wounds wet. Oh shit, no showers for two days. No! He grumbled constantly that I needed to receive the aid of a real hospital.

The doctor was angry to the point of swearing when he heard I had tried to jump from the banister. Leo ratted me out. He shook his head while saying he could give me no more of my prescription antipsychotics nor antidepressants.

He kept repeating that I needed professional treatment of my mental illness and the "serious" avolition wounds. I wanted to argue when Leo said that he agreed with the man, but I kept my word. I dared not complain for the sake of my lambs. Leo kept begging this doctor to tell Master Claus and Master Bladrick of his concerns and recommendations.

The doctor scoffed then said, "Forget it Leo. I am tired of speaking to those hardheaded fellows. I have told them repeatedly this boy is a danger to himself and others. They will not listen to me. I told his Master Peter before them too. I give up. I come to patch the mess they made of this young man. I keep my mouth shut these days. Not worth the chance that sooner or later one of them will grow tired of my nagging and exile my ass. They will kill him, or he will kill himself soon enough, then I won't be spending my days looking at the most revolting injuries I have ever seen done to any collar in this Haus. Good day to you Leo." He grabbed his medical bag and stormed from the apartment.

I let out a sigh of relief that at least I wasn't being sent off. Normally all I wanted was to be free of the Haus, but not anymore. I needed to break my collar and I had five little mouths to feed as well. I was a remarkably busy man without time to relax in some private hospital.

Leo sat down on the couch watching me pull on my breeches over the bandages. "I suppose you heard the doctor. No showers for two days and rest. Seeing that sandpaper business caused a lot of trouble my love. I want you to swear to me you never use it on your skin again."

I shook my head. "I cannot promise you that Leo. I may have to get more of it without getting showers for two days. The disease will be heavy in my flesh by then. It will outrun me if I don't take serious actions to stop it."

He frowned. "My love, you must believe me when I say there is no disease clinging to your legs. This is a figment of your imagination, a hallucination. Tell me why would I lie to you about this? Don't you trust me when I say I would never do a thing again to hurt you. I gave my word on my honor."

I thought about this for a moment. "I think you wouldn't lie to me on purpose, Leo. I wonder though if maybe you don't possess the proper power to see the truth all the time? There is only one of you. I have six. With more eyes, I can see many things you miss, ja?" That made sense to me at the time.

Leo looked at the floor. "Okay, I won't push this further. For now, I want you to promise no showers for two days, and no sandpapering in my home. You promise that, then I take you for another adventure this minute, if you feel up to it?"

I narrowed my eyes. "Depends Leo. I liked the lambs and the playground, but I don't desire to see more basketball nor the black collar brutes of this Haus."

He nodded. "Ja, I figured this out. You need socialization but it is not possible with your symptoms so heavy and burdens so rough. I think the chance of kids your age being cruel, to one that can take no more of that, is very likely. Therefore, it is a situation of rock and a hard place. You need to have interactions with complete acceptance of you from others. It must be positive and without stress. That sort of thing is almost impossible to find except with one type of friend. You give me your word you'll honor my requests then I will introduce you to a buddy that will make you happier than you could ever dream."

I admit I was curious about this fellow Leo was saying could be my best friend. I told you I am terrible with my interactions with others. I am sure even if I had met you under different circumstances Meine Liebe, you would not have cared for my company.

"I gave Leo my word, then he took up my leash once more. He led me out the door into the yard for the second time that day. I noticed that Vilber and Olaf were slipping me hateful looks when we left this time. I glared back at them without fear. Their clock was ticking down. I will never forget a promise. These two brutes were on my shit list for good.

Leo led me past the playground. I wanted to go swing but he told me that would have to wait till another time. I was disappointed but still eager to meet this boy Leo told me would be my pal for all time.

I started figuring out that something was very wrong with this unknown boy. Not only was he not in the Haus with everyone else, but they kept him in a barracks far from even the stables. He wouldn't be as capable of making fun of my nasty existence when apparently, he was even more disgusting than myself. I felt a little better knowing that this boy was obviously not the normal.

Leo took me through the door of a small stone cottage that had no glass in the windows. I became nervous due to the condition this fella was living in. There was straw all over the earth floor and no apparent type of heat anywhere to be seen.

Then I heard the soft sounds of whimpering. I stopped dead in my tracks backing up in terror when I heard that. I looked wildly at the walls noticing many chain leashes and leather collars hanging everywhere. Shit, this boy was so tortured he was mad and required being kept on a leash.

I wanted nothing to do with a crazy person, which was for damned sure. Leo didn't notice my fearful pause till the leash went taunt. He turned around seeming surprised at me refusal to follow him further into that cottage.

"Christian Axel? What is the matter, my love? Come with me to meet your friend. I had to go through a lot of trouble to arrange this. You won't disappoint him by acting the fool, will you?" He looked at me with curiosity at my fearful expression.

I shook my head, turning to head back for the main door "Nein, I am not so eager to meet insane boys, Master. I thank you for the thought, but if this is your idea of someone that would be my friend in a positive way, forget it. I rather be alone." I hit the end of my leash surprised at the strength of Leo when he held me from my retreat with it.

"You stop right there, my love. You come with me. This boy is not insane. You misunderstand. I would not dare to pair up the mentally ill. That would be a disaster. You come with me and for once trust my judgement without all this assumptions business." He pulled my leash roughly demanding I follow.

I groaned but minded when he reminded me my lambs where to be visited next for their dinner. I walked behind him listening with fear as that sound of crying got louder. I practically crawled up Leo's back in terror that this boy was indeed disturbed given the noises he made, no matter what he tried to tell me.

Leo giggled at my fearful behavior as he led me through a small entry room then to a large empty main room in that cottage. He stopped then pointed to the back of it telling me to set my eyes on the friend that would never betray me.

In the very back of the main room in a nest of hay, were six black puppies calling out with anguish for their mother. I gasped at the sight of them. I rushed forward forgetting myself and the leash that Leo held me on. He had to pull hard to stop me from getting into the middle of those baby hounds.

"Whoa, Christian Axel. We must speak a moment before you go over there. I have arranged that one of them is to be yours. We will take the puppy home after he is released from his nursing in three weeks. I wish this hound to be a gift to you my love, but I must have something from you in return for it." He had a little trouble getting me to pay attention to him because I wanted to hold the puppies too badly to listen.

I nodded. "Ja, whatever you want Master, I will give you for this hound. I thank you for the mercy. Can I go now?" I pulled harder on my leash whimpering almost as bad as the little pups.

He laughed hard as he rocked with the force of my trying to get closer to the yapping puppies. "You didn't hear what it is that I require in payment for this dog, Christian Axel. Do you not wish to know the price?"

I stopped pulling with a sudden fear coming over me. "Huh? Wait, you wish me to comply with your lustful interests? That is what this is all about? I lay there for your penetration and don't cry then, for this I get a puppy?" I frowned and looked at the floor realizing I had been expecting this shit from the beginning. No one is ever nice to me without wanting to fuck me sooner or later. ja?

Leo's eyes went wide. "Oh, meine Gott, nein. Shit, this is like the fucking broken record with you love. I told you I would never make you do that with me. In fact, now that I know you, really know you, I wonder how I could ever have thought of sleeping with you in the first place."

I winced. "I understand that, Master. I apologize for being the disappointment. I am not a sexual artist. I am the whore without feeling. Though you swore to love me anyway for payment. You cannot break that oath." I didn't look up at him feeling extremely low that he was disgusted with me, though I wasn't upset he never wanted to have intercourse. That was a good thing.

He sighed then covered his eyes with his free hand appearing fatigued. "Nein, I didn't mean it the way you take it my love. I meant that I love you, really love the real

Christian Axel. I understand now what you told me about everyone projecting their dark desires on a boy that is not interested in them. I can see I was just like that. I fantasized all kinds of things that I know are not truth. I wanted to believe them, but they were illusions. You are a fantastic, beautiful, strong, intelligent, loving, and generous person so it is easy to love you. However, you are also broken-hearted, tortured, lonely, and above all a child. I see only the abused boy that does what he must to live. Your mind is stalled further back then even your flesh. I watched you these two days and my heart aches from the devastation caused by Peter's little plan. I cannot unknow the pain you endure at the hands of those the view you as nothing more than a lamb in a cage grown only to fill their appetites. You are many things my love but a disappointment or whore, nein, never."

I shrugged. "Then if not demanding my special services, what is the payment for the hound Master?" I didn't believe a fucking word he just said. For the record, I knew better.

Leo looked at the pups that were practically howling in eagerness for my petting by this point. "I want you to stop treating me like your Dominant, Christian Axel. I want you to treat me like a friend or brother. I want you to view the apartment as your truthful home where no one can tell you what to do. No more kneeling, asking my permission, nor fearing telling me the truth. I demand you still follow only my three directives. No other will ever be given by me. You do that, then I will let you have a puppy. The puppy of your choice can come live with us in our apartment. When you come home you will have your family waiting with their tails wagging in happiness. When you must be away, I will attend to our baby. Promise me this, and then go pick our pet."

I looked up at him confused as hell. "I don't understand this thing you ask me, Master. You wish me to call your apartment my home, you and the hound my family, and never think of you as my Master again? Is that even legal?"

Leo broke into laughter causing me to startle a bit. "Damn, this is fucking harder than I ever imagined. Okay let me make this simple my love. Ja, I want you to break the law and be insolent to me. You can jump on the bed, eat all the snacks without asking, and even piss on the furniture if that is what pleases you. I will vow to never thud nor punish any behaviors in our home other than refusal to eat, take your medication or harm yourself in some way. This you understand?"

I shook my head. "If that is what pleases you Master, then I can do it I think?" I was truly thrown for the loop on this one. Leo had one weird fetish indeed.

He smiled. "Close enough my love. I will help you learn to understand this price over time. Go pick our baby hound." He let go of my leash.

I turned and ran to the squealing pile of fur in the hay. The hounds surrounded me as I gloried in their kisses and velvet paws. They all struggled hard to be in my favor but one of them caught my attention more than any of them. He didn't fight with the others. This pup crawled onto my lap and curled up. I picked up all his brothers and sisters enjoying their loving while he slept quietly.

I noticed that unlike all them he found his peace when close to me. I knew right away; he was the friend I had been seeking all my life. The soul that would never betray, judge, nor leave me for another. This precious pup was the one that could scare away my nightly visits from the boogie man.

With much gentleness I picked him up. He yawned wide and looked at me trying to reach my face with his pink tongue. I smiled at him, and he let out a loud jowl. I saw Leo standing there watching me look him over.

"This the one, my love? He is lovely. Pure black. Even his eyes. Look at his stub tail wagging with vigor. I like this pup." He knelt to pet the baby hound.

I nodded. "Ja. This is my hound. I will name him Der Makellos (the unblemished). When did you say he can come with us?" My new friend licked my nose making me and Leo giggle.

"Three weeks my love. Der Makellos needs his mother a bit longer to grow strong and healthy. You would not deny him such a good start, ja?" He took the pup from my hands snuggling him to his neck while the hound tried to lick him everywhere.

I laughed. "Nein, I would not Master. He likes you. See he kisses like the perfect pleasure submissive."

Leo snorted. "I agree my love. Hell, with this much adoration who needs a lover, ja? I should have gotten a hound ages ago. I would have been a less grouchy old man." We both laughed for many moments at that.

We spent another hour with Der Makellos, then went back to feed my lambs. When the task was finished Leo took me with him to stroll the gardens. He was friendly with all the Dominants we encountered. I knelt while he spoke to them every time the way I was trained to do. He seemed irritated by it, but he knew if I had behaved otherwise, it would cause tongues to wag.

When the day grew long, he took me back to the stables and I attended my lambs dinner. They knew I would feed them by now. It was a pure pleasure to be so damned welcomed by their eager faces. Life was finally peaceful, stress free, and quiet. I could almost forget the horrors of my existence on the loving leash of my Master Leo.

We returned to the apartment. He played his records and read his romance novel while I worked on my studies. He had me eat the meal he called in for and I wondered why he didn't take me to the Great Hall to show off his possession. I wanted to ask him but decided it was not my business. He got on to me politely every time I knelt, asked permission to make water, or get anything from the kitchenette.

I was having trouble with this complex concept that his home was my own, for truth not just his saying to make myself at home either. No matter what he said I continued to behave as the submissive I was taught to be. He was my Master and my better. I couldn't seem to stop my kneeling, fear of his anger nor constant hesitation to engage him in conversations he didn't initiate first. Believe me, it was damned quiet unless he took up that slack.

That night when he went to bed, I checked my legs for the illness. To my shock, it had not returned. I thought of that sandpaper cure. It seemed to me no matter what Leo said that shit worked. I decided to never break any vows to Leo, but the minute I went back to Master Jonas, I was getting more of it. I knew that disease was coming back, and when it did, Mad Max would be ready for it.

The next day was peaceful as the they before. Leo took me to attend my lambs and visit with our pup Makellos. He told me many silly jokes and took me to the swings. We strolled the garden again, and from a safe distance even watched the basketball game of Edger and his black collars.

I was having the time of my life. I thought if everything could be like this all the time, I would have no further quarrel with the Haus nor my collar. It seemed to me Leo was the best thing that ever happened to me. If I ever loved him before, now I was beyond enamored. He was fast becoming the only thing that mattered to me in all the world.

Master Bladrick and Master Claus showed up looking for their leash rights right after we returned to the apartment that night. I was a bit upset but I recalled the deal I had made with them to get Master Claus to relent on his order to keep me from my rounds. A promise is a promise, and I am a man of my word.

Leo argued with them that I was not healed from my vigorous scouring with that sandpaper. The Elders were becoming heated in their discussion. I motioned for Leo to call me aside. He came to me, and I told him I would go with them to keep the peace. He was most unhappy, but I reminded him that discord would injure my chances to break my metal. That truth forced him to release his hold on my leash to the lusty old schwulers.

True to his word Mad Maxx took the wheel solitary without the ailing Maximillian. For two days the poor boy had to put up with the London Bridge and tolerate the octopus hands of those two old perverts. I took my rest but kept my eyes on the masochist in case this foul touching shit got to him deeply the way it did our submissive brooder.

To my relief Mad Maxx handled it with honor that made me gain a lot of respect for him. He was not allowed to remove his dressings though. That made his showers impossible. I could see the disease starting to take hold again.

I witnessed this was starting to bother my strong brother shard. I began to worry when by day two, he was sweating often, and the boy cried during one of his special services requests with the old men. I was glad that this nightmare was nearly over. I didn't know how much longer Mad Maxx could hold out.

The real terror began when the Elders refused to relent their hold on the boy when Leo came to pick him up at midnight, the second day when normally their cycle would be over. Master Claus argued that they gave up their two days on the last rotation and now demanded they be made up.

I saw the look of despair in Mad Maxx's eyes when Leo was outvoted. Mad Maxx had to stay with those dirty old men another two days. I rushed to see the masochist when he went to the washroom to scrub his hands after Master Claus and Master Bladrick sent the angered Leo away.

"Brother how are you holding up," I asked with anxiety at hearing his answer.

He shook his head. "I don't know Mad Max. This shit is getting to me. I never realized how bad Maximillian had it before. I mean I have been here riding the flesh with him but never handling it all alone. I think I may come unglued if I have to swallow another mouthful of bloody piss, I swear it."

I winced at that. "Is Maximillian rested enough to aid you brother? What does Der Hund say?"

He looked at the floor. "Der Hund said nein on calling in Maximillian, brother. He is sick, very sick. This sex with the man finally got to him. I don't blame him either. I think jumping from the banister may be the right answer after all."

My eyes went wide. "Nein, brother you cannot do that. What will happen to the lambs and Der Makellos? They need us. You must deal with this."

Mad Maxx sighed. "I will not do any of those things, Mad Max. You are the one that was honored to hold such a joy. For me and the submissive, there is only servitude and laying there used like the sex doll."

I moaned out in sheer fright. "Ficken mich, move over brother. Max says Der Hund has ordered I partner with you until Maximillian is healed. This is bullshit for the record."

Mad Maxx smiled with bitterness. "Does this mean I get to listen to the records and pet the lambs, too? Or do I get sent to the nothing the second this horror show is completed?"

Max with his goofy grin chimed in. "You are most welcome to stay when Leo begins his next clock brother. You need the pleasure as much as the pain as you have tolerated. We shall all share the tears and the laughs, ja?"

I wanted to beat Mad Maxx and Max to a blood spot in the tapestry, but I knew one would enjoy it. The other would likely destroy me before I could finish the task. I was stuck, Meine Liebe. I jumped into the boy joining our ailing brother.

I will not bore you with the details of the next two days of this indignity. All I will say is that I hate London and I hope all their bridges burn the fuck up. When we weren't tolerating their groping, foul kissing, or that nasty threesome business, we fulfilled every other service a pleasure submissive can attend.

We did bath service, wake up, message, conversation, entertainment, and every other service for two. It was not only a foul task, but it kept us so busy we had no time for anything else. It was like taking care of babies that were bigger, heavier, and more demanding than the real ones. Throughout the whole thing, Mad Maxx and I took turns enduring the torture of it.

Master Claus's gout was acting up and Master Bladrick was having difficulty with an old knee injury. They never left the apartment the whole four days. Strangely, they never allowed us to leave either. They called in all meals and had Leo come to give medication without having us go to him. We wondered if the Elders were suspicious of the rapes or if Leo told them not to let us leave.

Whatever the reason, it was four days of oppressive, crushing subjugation that I had never hated more. It was not that Master Claus and Master Bladrick were cruel. They were kindly but demanding and touchy feely.

However, now that I had a taste of life without the enforced services and treatment of being nothing more than a cock holder, I appreciated what Leo was trying to teach me. I just wanted to go home. I missed my lambs, Der Makellos, and above all I missed Leo. The terror of being a pleasure submissive had become more burdensome than ever.

On the third they I saw the disease rising to levels of danger. Mad Maxx and I ripped off the dressings and started our schedule of scalding showers. It was difficult to work them into the many demands of our Masters. Though we didn't have the sandpaper we made that rough rag count. The disease was eradicated when we managed to get ten cleansing in before on the fourth day, Leo's clock began at midnight.

I was thrilled to see Leo there at the door with his beautiful smile to pick me up. Mad Maxx was still trembling from the horror of the last four days. He perked up when that man was spotted waiting for us. He shot me a smile of pure joy.

All he had done for two days was beg us to tell him repeatedly about the feeling of the coats of our lambs and the softness of Der Makellos eyes in the morning light. It was his desire for the experiencing These things for himself that got him threw the roughest moments of this horror life we endured.

I practically knocked Leo over with my kisses of gratitude when the Elder Claus closed his door leaving me alone with him in the hallway. Leo laughed with immense joy for a moment. Then he pulled me off saying that I reminded him of Der Makellos with my overeager adoration of him. I chuckled at that comparison telling him that was an honor I was happy to accept.

We returned to his apartment and as usual he wanted to go to bed for his rest. To his surprise I followed him. I climbed into his bed without removing my clothing and fell asleep in his arms. I didn't even take a shower. Mad Maxx and I were whipped. We needed the rest after that nightmare. Besides, for the first time in our life we finally understood what the word "home" meant. With Leo we had found one at last.

The next four days brought us peace and happiness once more. Mad Maxx adored our lambs and hund with wonder that even made me, the sadist, smile. When Leo was not taking us on long walks, conversing with us or making us eat, he was reading his novels, or reminding us that his Haus was our own.

I was in seventh heaven forgetting all about Master Jonas, and the London Bridge Masters. In the haus that I now called my home, I could even pretend that shit wasn't even real. Yet, every day when Leo went to nap or left for an errand, Mad Maxx and I would hit the shower. The disease was back with a vengeance. I feared at the rate it was spreading it would soon overtake the flesh.

Mad Maxx and Max had no clue how to stop this rapid growth. It was beginning to scare me a great deal by the fourth day. I started to wonder where Leo hid the sandpaper. Max reminded me that we had gave our word to never use that tool in our home. I was not sure what to do to fix it when I spotted a wire brush under Leo's sink. When he rushed out to get another romance novel, I put it to use.

I was excited to have found this cure. I could barely wait for the water to heat up before rushing in to try out this tool. My bruises and cuts from that horrid rape had turned to deep browns and yellows. The last scraping of our flesh had begun to heal as well. I realized thanks to all that it would take more force to remove the boy's hide this time.

I had only just started on my left leg when Leo came rushing into the washroom. I was surprised to see him through the shower curtain. He had been giving me my space and privacy as promised. I didn't understand why he would invade my personal hygiene time like this.

I tried to hide the wire brush when he snatched back the curtain glaring at me with anger. "Christian Axel, you come out of that water this minute. We had an agreement, remember? Only one shower a day. This is the second one since this morning. Get out now."

I shook my head looking at the drain watching the fresh blood roll down it. "Nein, Leo. This is the only one today. I don't recall another one." I was grateful he had not seen the brush.

He shook his head. "You may not recall it but there was the one. You want to argue this fine. Get out and do it with me where I will not be burned to a crisp by this boiling stream you insist to shower in." He stepped back pointing to the floor in front of the tub.

I groaned but turned off the facet and stepped out trying to keep the wire tool behind my back. Leo saw that I was hiding something. He tried to get a better look at it. Then I saw his eyes get with in horror.

"My love, your left leg is bleeding badly. What have you used to do this kind of damage? Let me have whatever you have behind your back. I will only ask this once. If you refuse me then I will leave this room and let you fend for yourself. This is a small wound, but I think if I leave, you're likely to rip yourself to pieces with it. Then you will die, and so will the lambs, maybe even Makellos and me from the grief. I beg you to recall your promise to me. No harming yourself in this haus." He held out his hand.

I winced but handed the wire brush to Leo. His eyes focused on the thing for a moment. He dropped his gaze to the floor with an expression of extreme sadness. Then without another word he left the room taking the brush with him. I stood there shivering unsure what to do. I had obviously upset him a great deal but even giving him what he asked didn't seem to make him less disrupted.

I shot a look of terror at Mad Maxx "What the hell? Leo is not happy with us, brothers. He seems to be anxious to keep us from treating our disease. I thought he is our friend with all the kindness and freedom he grants us, but then there is this dishonor of taking our tools."

Mad Maxx looked at me with an expression of disbelief. "Mad Max you are the Gott damned fool. Can you not see the problem here is that you deny him the special services he has earned. That is why he is really displeased with us."

I shook my head. "Nein, Leo says he doesn't desire that from us not ever. He swore to love us without such that burden."

Mad Maxx crossed his arms with a bitter grin "And you believe that bullshit? Surely even you are not that stupid, Mad Max. This man is a schwuler that confesses that he wanted to fuck the boy. Now you think the Master will just be satisfied without this service. Look at the expense he hands over on our behalf. He gives us everything and what do we give in return for all of it? Not a fucking thing. Instead, he has to watch these other foul men get what he is denied. That is why he is keeping us from this cure. Equal for equal, brother. We don't give him any relief, so he takes our own, ja? Did you learn nothing over the last two years in this collar?"

I dropped my head realizing the masochist was right. "Shit, I never thought of it that way. Yet, the other day he told me he didn't ever want to sleep with the boy. He said we are too familiar to him, and this bothers him."

Mad Maxx scoffed. "I heard that horse shit conversation too, brother. I will tell you what I heard. He said he will never request the special services. He said he will not make the first move. This is just one of the games the Dominants play with our head. Think Mad Max. Like Jonas, he wants us to prove our love for him by offering without his having to demand them."

I nodded. "Ja, which makes sense I suppose. Does make me wish that he would just be honest and say this to us though. Why must they all be so fake? I really thought this man loved us. I really did."

Mad Maxx sneered. "Like maybe Ryker did? How about Geraldine? Ah, I know like our father and mother do. Nah, this one he loves us like Annette does."

I looked up at him with a startle. "Annette does love us for truth brother. I can agree with you on the others but with her it is different."

He giggled. "You are daft, Mad Max. Where the fuck is the police shutting down this Haus? You gave her the door to escape and run so she did. She kept going too. She forgot all about her precious Mad Max left here to suffer the dishonor he does, didn't she."

I shook my head feeling I may cry at the painful truth. "She did tell the police I know she did. They didn't believe her. Annette does love me. Besides, I made her promise to forget us brother. She is a woman of her word, is all."

Mad Maxx nodded. "You keep telling yourself that. I for one will prepare the boy for the penetration sex with Leo. You will see that if we don't seduce him and keep him happy, he will tire of our antics and expense. Then we will be forgotten like we always are. Didn't you say to Maximillian and me that you would be willing to endure his intercourse to keep him for your own? Well, if you won't, then I will. I wish to keep his affections."

I put up my hand demanding his silence. "Fair enough brother. You need not do this. Leo is my responsibility, Der

Hund said so. I will handle this indignity and humiliation. Then he will give us back our tools and we get to keep the lambs with our hund."

Mad Maxx smiled with bitterness. "Cheer up, brother. At least we love this one, ja? He did promise the blow jobs sometimes. It won't be so bad. You'll see."

I sniffed back my tears of disappointment. "Ja, you are right. It will be tolerable. Maybe we can close our eyes and imagine that we are in the fields with our pretty lambs and Annette too. Then when it is done, we kill the disease. One day we break this collar and run away with Der Makellos and our sheep."

He nodded. "Now you are speaking my language, brother. Get the boy ready. Do not fail at this seduction. We don't have much time before the Vampire gets home. If Leo is this despondent when he returns, maybe next cycle he will forfeit his leash with us."

That possibility put a bee in my bonnet to get to work with the prep immediately. I feared that Master Jonas would eventually bully Master Leo from his time with me. If I offered the coupling, then Master Leo may stick around awhile longer. I had noticed over the last few days he was often irritated by me, and more silent than usual.

I suppose I had already figured out this sexless love life with me would never work out. I don't know why I hadn't already offered it with vigor. Not like one more bed mattered to a whore like me anyway.

When I was ready for his intercourse, I dressed in a simple blouse and breeches. I didn't bother with boots or socks. I was ready to give this Master his rights without quarrel or hesitation. I walked into the living room to find Leo on the couch weeping into his hands.

Seeing him so grief stricken frightened me. I rushed to kneel before him ready with my apologies for whatever I had done to make him cry like that. He startled when he saw me at his feet.

"Christian Axel? What the hell? I told you no more kneeling. Damn it. No matter what I do I am not getting through to you. I swear I am ready to give up." He wailed out.

My heart stopped beating in my chest as my blood ran cold. "Nein, I apologize Leo. I will try to please you harder. I understand what the matter is. I am here to fix this. Please I beg you let me make this right." I leaned in pulling his head into a deep passion filled kiss.

He sniffed but kissed me back. His muscles were rigid and full of stress initially. As I kissed him deeper and then moved to his neck and ears, he shivered and moaned, filling him with passion. I braced myself to take a deep breath to strengthen my resolve as I pulled off his t-shirt.

Leo didn't give me resistance as I kissed and licked down his chest. I reached out and undid his jeans, finding his interest full of the hardness of the oak. Without hesitation nor thinking I dropped my head into his lap and began my oral stimulation of his manhood.

As twice before, he began to vail out and pant over my skills. I didn't find this humorous as I did before. The further I got into this early intercourse the more I realized Mad Maxx was correct. Leo was interested in intercourse with me despite what he told me. He was better than the others only in the amount he was willing to pay to have me on my knees. Oh well, sucks to be me, ja?

Before Leo could spend himself, I pulled away from my task of giving him head. I went back to kissing the now very lust driven Dominant. When I was sure he could take no more of the teasing I reached down and pulled down my breeches without breaking from my lustful kissing.

I pulled away removing them entirely. Without a word I took the position of submission. I patiently awaited his couple with slow shallow breathing like I always did for this penetration business. I hoped he thought to get his lubrication, but I was not going to complain no matter what he did or didn't do to me. I closed my eyes and thought of my lambs and of Der Makellos. I hoped that Leo was quick with reaching his satisfaction so I could get my tools back. I already felt the need for the showering.

To my shock Leo spoke out in a raspy voice, "Christian Axel, get up from their and stop this nonsense. I do not want you to be like this, I told you that at least a dozen times already. This is hurting my feelings and cannot be anything but a humiliation for you. I refuse to be seduced by things that are false. My eyes are no longer blind to the truth of you. I should have put a stop to it sooner, damn me. What the hell was I thinking."

I picked my head off the floor to look back at him. "You don't wish for this service? Leo, I am willing to do this. I won't even cry I promise."

His eyes went wide and filled with tears. "What? You won't even, oh dear Gott. I cannot handle this, Christian. Put your clothes back on. I apologize for misleading you, for using you, for abusing you, for everything. I am so fucking sorry for what I have done to you I cannot event stand to look at you anymore. Gott damn me. The shame of it is killing me." He got up from the couch and stormed down his hallway to his room slamming the door behind him.

Mad Maxx and Max stared at me in complete terror. I sat up looking at the wall with his words ringing in my ears. All I could hear was that he couldn't stand to look at me anymore. That my foulness was killing him.

Oh, Meine Liebe, you understand how this made me die inside, to hear the one I loved to say that to me. I couldn't find the strength to stop the nasty touching of the other Dominants or the rapist. I was a nothing that wore a manacle of silver around my neck. Leo deserved much better than this scarred up, worthless, and used boy.

I was told that I am the strongest shard because as a sadist my heart cannot break. That was a lie. That night it did. The pain of it was so deep, I could not even force the tears of it out for any relief. I stood up without redressing and went to the washroom in a trance.

I knew the problem was that fucking disease of my flesh. Leo didn't want to become infected with it. This was understandable. It is the foulest of things to have clinging to you. I went to my bag and dug through it till I found my razor. It was time to remove all of it once and for all. Then when I was down to the bare bones, I was sure Leo would love me at last.

I turned on the hot water and stood there waiting for it to heat to the scalding temperature. I braced for the agony, then stepped inside the tub. The door flew open causing me to nearly fall down with my flinching at this sudden movement.

Leo came across the room and grabbed my wrist wrestling the razor from my hand. I was stupid enough to carry it in my left hand and my broken finger on it was still weak, making his task easy enough. I wailed and fought him trying to keep him from taking my final option tool. He was assuring me I would be eaten alive by the sickness. I didn't know why he was tormenting me.

"Nein, don't take it away, Master. I need that and the showers. I need them to be well. Can you not understand? If you allow me to have it back, I can be clean enough for you to me. If you leave me be, I can fix your disgust over this foul flesh. I apologize for being undesirable for you. I only wish to please you. I should have known better and done this a long time again. Forgive my weakness, I beg of you." I fell at his feet wailing in the prostrated position full of hopelessness that nothing I did ever seemed to make him as happy as he made me.

Leo threw my razor across the room then reached out and caught my leash holding me back from retrieving it. "Nein, my love, halt. Come to me know. Kneel, damn you."

I looked back at the razor laying in the corner but dropped to my knees before the breathless Leo still holding my leash. "Please Master, mercy. I need that and the water to be clean for you. I only wish to make your heart sing like you do my own. I wish to serve you. I beg you to allow me to. I love you. I cannot bear to lose you."

Leo reached out and pulled my head to his chest weeping loudly. "Honey, you will never lose me. I swore that to you. Why can you not hear me?"

I wailed with terror. "I hear you, Master. Please, I need the water. It is the foulness that keeps us apart. I know I disgust you. It disgusts me. Let me show you I can grow a new boy. You will want him. I know you will."

He sobbed and held me rocking. I cried for a long time, unable to calm down from my anxiety that he would eventually leave me like the few that cared did. I saw Ryker's face broken, and the tears in Annette's eyes as she walked out my door. I couldn't take any more of this horrible loneliness I had endured for so long. I had to keep Leo with me no matter what I had to do to keep him.

Leo finally talked me into coming to his bed for a cuddle. He insisted we keep our clothing on and thwarted several of my attempts to engage him in intercourse with me. When I would begin to paw him, he would gently pull my hands away. I was more confused than ever and fell asleep more depressed than I had ever been in my life.

Leo would tell me a few years later that after I cried myself to sleep in his arms, he lay awake. He was unable to find his slumber thanks to the heavy burden of his insight of my painful plight. Leo knew this terrible problem was one he helped to create and the shame of it brought him great dishonor.

He did love me with all his heart. Leo's affection for me was so pure he couldn't talk himself into engaging in a couple with me. He knew I would find nothing but humiliation and self-loathing from it. He understood that my constant showering and scouring was my way of combating the horror that enforced sexual contact made me feel inside.

Yet to not couple with me brought the same level of terror. I had been subjugated and raped to the point of not understanding that sex was not all I was worth to him. His refusal of my advances brought on a worse reaction than if he had just done the dirty deed. He was in a bad spot with two bad choices.

If he coupled, then he would think less of himself because he would know he forced it by proxy. Then he'd have to watch me slowly scour my flesh to the bone. I would likely die of infection or cutting too deep then bleeding to death from it. If he refused, then I became suicidal believing that he was no longer interested in keeping me around.

The clever Leo had already found a way to keep the lesser suicidal wishes down by giving me something to live for. The lambs and my hund needed their Christian Axel or they too would find a place in the orchard. He used my love for them and my natural need to defend the helpless to keep me going when I was out of gas and running on fumes.

He found out quickly that was only the tip of the iceberg in assuring my continued fight for survival. His biggest problem was finding a way to convince me that he was my lover and friend that I could trust. I needed that more than anything else in the world at that moment or there was no doubt very shortly, Christian Axel, the Mad Maxx, was to be no more.

That night, as he held me in his sweet embrace, he realized the answer to the problem that none of us had been able to solve in all the years we had worn our collars. He said he prayed to his Gott that his plan would work to bring a lost frightened boy out of the darkness into the light of his heart where he belonged.

The next morning, I felt empty and anxious. Leo was already up. He demanded I get ready and come with him to attend my lambs and our hund. I found much joy in both tasks but the second they were out of my sight; my darkness began to rise within, more vicious than ever. I was not even able to laugh at Leo's constant silly jokes and kid like antics.

I felt damned low that I could not even seduce the man that told me he loved me. If I couldn't bring him to desire my favors, then what the hell good was I as a pleasure submissive? Seemed every damned pervert in the world didn't have issue with soiling the fuck out of me, but not my Leo. He was completely turned off the second I tried to submit to his penetration. I was ready to take that flight off the banister and even Mad Maxx and Max agreed, we were Gott damned useless.

That day when I was not attending my animals or studying, I sat quietly on the floor staring at the floor, damning my existence. Leo read his novels letting out his gasps and grabbing his chest from time to time when he hit a very romantic part of it. He never left my side.

I ate his meals without interest or vigor and took the pills barely looking up at him when he gives them to me. He continued to exhibit this odd cheeriness. I wondered how he could be so damned thrilled when he had himself leashed to such a loser. I continued this constant self-berating for the rest of that afternoon and all the way to bedtime.

When eleven o'clock came I expected him to go to bed. I thought I would maybe pick his door locks and take that flightless plunge to the first floor. I was shocked when he took up my leash and told me to come with him but keep silent. I was thinking maybe he was so sick of looking at me he was turning his clock back to the Elder brothers Claus and Bladrick. I tried to go get my bag, but he held me tightly by that chain demanding with more force that I follow him.

I kept my eyes to the floor as we went down the stairs. I then began to think he was taking me to the torture rooms for a well-deserved thudding. When we passed up that stairwell I was really confused. I started to tremble thinking he intended to dump me in the dungeons.

Once again, he passed up that stairwell. I looked around trying to figure out where the hell he could be taking me at this late hour in the Haus. The hallways were mostly devoid of staff and silvers. The nightshift was working but since most of the residents were sleeping or settling down for their slumber we were not slowed down with greetings or kneeling collars.

Then he took a hallway I had never been down in all my time there. I looked around at the huge paintings of long dead people and fancy tapestries and colorful expensive rugs. He turned off to the right through a large arched entry.

To my amazement right there in the center of the room was a huge tub of water the color of the sky with lights under the liquid. I looked up to see the roof over it and walls all around. Leo smiled at my look of awe. He giggled as he told me this was the Haus's heated pool. I had seen such things in magazines, but I didn't know they were real things.

I pulled my leash to the furthest end of this marvelous creation. I saw him watching my look of fascination at this most magical appearing water. When we got to the end, he showed me there were steps that allowed a person to walk right into this pool. I was thrilled to see such innovation.

He asked me if I would like to get into the water with him. I nodded that I would love such a thing, but that I could not swim if it was very deep. He chucked with the idea the someone that loves water as much as I do could not even keep his head above it if got too deep. He told me to strip. and I did this without being told twice. I was eager to get into the soothing stuff and see if it worked like shower did at killing the illness.

Leo went in first and held my leash tightly telling me he would not be letting it go. He was not comfortable with my wandering around the water without him since I was unable to even float. I stepped into that heavenly pool and closed my eyes the second I was fully emerged to my waist.

I felt immediate relief of my pain like in the tub. I opened them and shot a smile at Leo. I could smell the chemicals in the air. Leo told me this was chlorine to keep the germs and staleness from the still water. I realized this disinfecting stuff was beating that clinging filths attempts to take ahold of me.

Leo saw my relieved grin as I splashed about in this wonderful cure. "Does this make you feel clean being in this special water, Christian Axel?"

I nodded. "Ja, I can stay in here for all my life. Then I never get sick again." I dropped down into the water submerging my head then popped up with the thrill of feeling fully new for the first time in many years.

Leo splashed me several times giggling like the schoolgirl. I splashed him back, laughing. Then I sat down, stuck my head under, and walked about in the "shallow end" of the pool. I was looking at the deep end longingly wishing I could swim when I felt Leo tug my leash gently.

"Come to me my love. I wish to kiss you in the pool, ja?" He grinned while I approached then engaged him in a playful kiss.

I went to pull away desiring to look around the pool some more when he grabbed my head pulling me back to his lips. Within minutes I realized his embrace was becoming very heated. I kissed back unsure what to make of his sudden interest in my adoration. I wondered if he was going to think I could give him a blow job under the water.

He appeared to hear my thoughts. "My love, just focus on me. I wish to accept your offer for a couple right here this minute in the water, but only if you can find your lust back with me. If you cannot find pleasure in my penetration, then the act is only a sacrifice for you and a theft from me. However, if you and I can enjoy this lustful act together, then I think you will understand there is a difference between sex and making love. Would you like to try this? If you wish to say nein then I will still love you for all time. There is no pressure, I swear this on my honor. I will wait a lifetime for you to be ready or even if you never are."

I nodded. "I wish to please you, Leo. Thank you for allowing me the mercy of it."

He stroked my cheek and kissed my forehead. "Nein, my love. It is the other way around this time. I wish to please you, Christian Axel. Thank you for allowing me the mercy of it. Do me the honor of alerting me if anything I do upsets you and I will reward you with my undying affection for all my days."

He then spun me around slowly pushing me to the corner of the pool. We engaged in ever increasing kissing, nipping, and licking of each other's neck, ears, and lips. I felt him reach down and gently stroke my manhood. I returned the action without hesitation or disgust. I shot a cat-like smile at him when he found me most interested in his skilled touch. I gained an erection as I had on his couch days before. He grinned back then increased his rubbing and touching of me everywhere.

He let out a gasp when I suddenly dropped under the water. The water lapped at Leo's flesh as I engaged him in aquatic oral sex. I returned to the surface to gasp in air then dropped back down to attend my lover's eager cock. I could hear him moaning out loudly in his ecstasy even when my ears were fully under the liquid.

I came up for air after several moments of this unusual oral service and he reached out taking me and my arms to prevent my going back under. He pulled me to his excited kissing. Then he turned me around, placing my arms on the steps. He reached out to grab his pants and took out a bottle of lubrication. I sighed with relief that he was granting such mercy as I endured his preparing for his mount behind me. I closed my eyes and braced for his entry.

He was gentle with his engagement, taking much time to assure I was not too harshly violated by his initial coupling. Leo began his thrust slowly and easily, moaning out in what sounded like pure thrill. I kept my eyes closed hoping that he would be well pleased, then never leave me. I thought of my lambs and my hund. Though I confess I was near tears already. No matter how much I loved Leo, this anal intercourse was not something I could endure well.

Then to my shock, he reached around my waist and took my own manhood into his hand. In a perfect rhythm he stroked me and thrust at the same time. I could no longer think of anything other than the electric feelings of lust burning into my teenaged brain. I began to pant out, unable to contain my own cries of excitement.

Leo was encouraged by my sudden calls of interest and the evidence of it growing ever stronger in his grip. He increased his efforts which sent me to near orgasm in short order. I yelled out that I could not hold on much longer, and he wailed back there was no need for me too. We found our apex at nearly the same moments, there in the water of that magical pool.

Leo had given me my first blow job, my first experience of loving another man, and now he granted me the mercy of finding my first orgasm during a couple with one of my own gender. He uncoupled after panting and yelling in happiness for a few moments. I couldn't catch my breath either.

We embraced and kissed with slow passion that only lovers' can know after a satisfying climax with the other. He smiled at me as I hugged him tightly, feeling relief and true love by another at last. I was grateful that he chose the magic pool for this unnatural act, and waited till we could be alone.

Too bad we weren't alone. Someone had watched the entire show, and she had an axe to grind with Leo. Gretta could barely wait to share with all her friends the latest gossip she'd seen with her own eyes. The straight Priceless collar had found an orgasm in the couple with his Master Leo. That could only mean one thing. Mad Maxx was in love with the Elder Leo and Master Jonas was due back in only two more days.

Chapter 48: The Blue Water Cure

Leo sat on the steps of the pool and pulled me into his lap. He kissed my mouth with loving slow passion. I admit I was adoring his affections a great deal. He had lct go my leash and wrapped his hands around my lower back in a lover embrace. I put my own arms around his neck falling deeply under his spell of tender touching and romantic speaking.

Neither of us saw that rat Gretta slipping out of the swimming pool room thanks to our complete attention to our post-coital adorations. She had silently watched our passionate lovemaking, the affectionate aftercare, and then snuck away without detection. We thought ourselves in privacy the whole time. That would be a stupid mistake that was going to cost us both a great deal of personal agonies.

That moment though, nothing else mattered in the world but the joy I felt that Leo did love me. My seduction had been successful at last. I could let the worry that he would tire of holding my leash go for a while. That is until he tired of my artistry. That is the way of things. Nothing lasts forever, especially a Master's thrill at their submissive.

I pushed the memory of this heart-breaking truth from my mind. I worked hard to obtain Leo's favor. I decided to just be comforted, it was mine for now. I knew I had plenty of time to grieve over the loss of his love when the day came, he no longer desired my company. I broke from his kissing to snuggle my face into his neck. Leo dropped his chin onto my head and stroked my back with much gentleness.

He held me like this saying nothing for several minutes. He seemed to be unwilling for this peaceful time to move forward. I listened to his heartbeat with thoughts of the lambs and our Hund Der Makellos. I wished with much silliness that Master Jonas would never return, and that Leo could have my collar for all his own. I had wanted a Mistress with all my soul, but I could be content to have a Master if it was this man.

Leo cleared his throat pulling me back and looking into my face, "Meine Liebling Spatz (my darling sparrow, a common German pet name for a beloved family member), do you feel unclean? Do you see the illness clinging?" He smiled bitterly.

I looked down at my legs and smiled. "Nein, Leo. The magic water is killing the disease before it can take hold. The liquid is the cure. You believe me, don't you?" I looked back at him searching his eyes to see if he understood.

He nodded with an expression of seriousness. "Ja, I do meine Spatz. I brought you here for this special treatment. This pool has the power to take all the pain of the past away. You can feel it inside, ja? It is truly clear that you are not the same Christian Axel that walked into this water. I see a new boy in my arms. This is the one I love with all I am. I will never allow another in my heart for all my days. I am reborn with him, and the bond of our mutual regeneration is unbreakable."

My eyes went wide in shock. "I thought this water was supernatural. There is none so blue nor warm in nature. I do feel that I am not the same anymore as I was before this treatment. What is the magic in this fluid, Leo?" I cupped some of that amazing water trying to get a better look at it.

He chuckled. "Do you think they would tell Leo their secrets of chicanery? Nein, I have asked for many years. I even offered them enough money to buy their own island, but they deny me the answer to this day." He furrowed his brow as if upset that these, whoever they were, would not share their tricks with him.

I craned my head around to look for these people with such amazing powers of healing. "They are likely watching us, Leo. I am not sure they will appreciate that we soiled their perfect vision with our perversions." I frowned as I thought of some strange people that were probably pretty pissed off at us for coupling in their mystical water.

Leo laughed loudly. "Ah, Meine Hase (my bunny), they made this pool for lovers coupling and for soothing minor aches and pains. This thing we did was no perversion, Christian Axel."

I looked back into the blue reflection of myself staring back at me. "As you say, Leo." I knew better than that lie he was telling.

He frowned at that. "Christian Axel, there is no shame in finding pleasure when making love. The ability to join with one you adore is a gift most special. I want you to know, I will appreciate what you sacrifice to demonstrate the depth of your truest affection for me. I will never take that for granted, nor forget that you would never have been with me in a million years had it not been for ill fate. I can never thank you enough for caring for me like you do. Ich werde dich für immer lieben meine Liebling." He kissed my mouth gently after his vow to love me for all time.

I kissed him back and wanted badly to believe in his words, but to be honest, I didn't. I refused to be betrayed yet another time. I merely repeated that I felt the same way about him. That was truthful, despite the fact that I expected Leo to use me and throw me away at some point.

Leo smiled with humor. "Come with me now meine Hase. We need to dry off and head home. The hour is late, and the lambs need breakfast early. Besides, I am more wrinkled now than ever in this water. I get anymore cracks in me you will run off with a SharpHund thinking it was your Leo." He took up my leash easing me off his lap.

I followed him up the steps sad to be leaving this marvelous pool. Leo saw my expression of longing as I dressed while staring at the water in a trance.

He shook his head with a bit of sadness in his expression. "Christian Axel, if you will stop scouring off all your flesh and cut back taking so many showers, I will hire a black collar to teach you to swim. They give the lessons in this very pool you know. Do you think you can do that for me for this gift?"

I broke from my trance. "Thank you for the mercy, Leo, but nein. I cannot end that battle without being overtaken by the disease. I will not make a vow I will never keep. I will have to learn to swim when I get away from this horror hell hole."

Leo looked at the ground, his expression full of concern. "This scouring you do to yourself will never stop, will it? Is their nothing I can tempt you with to stop this mad behavior?"

I startled. "Mad? Nein, Leo. It is not crazy at all. I grow back. There is no cause for your concern. I make a new boy each time. You will see. I grow the new flesh but first I removed the sick stuff. When I get out of my collar I will go to the medical school, and they will teach me how to cure the return of this foul disease. Until then I keep it from consuming me anyway I can. If you are well pleased, then I respectfully request you return my sandpaper and razor. Thank you for the mercy of it, Master." I smiled with happiness that I was able to earn back my stuff.

He shook his head. "Nein, I will never give you back that sandpaper or the razor. You thought I took them because you were not providing the special services. Oh shit, this is a fucking nightmare. Okay, I try another route to get you to listen to me. This brainwashing has made you deaf to all reason. I didn't want to go there, but I must speak your language, or I get ignored. Christian Axel you listen to your Master. I find this scoured flesh ugly to my sight. I do not want my submissive to attend to my services looking like a side of raw meat. I bet you can hear me now, ja?"

I glared at Leo with anger. "Then I suggest you put on a blindfold when you fuck me. No one seems to have issues with stripes, bruises, welts, and stitches. Well, the Master can overlook my cure wounds and always have. I serve you, Leo, anyway and anytime you desire. Dress me up, cut my hair, make me walk like I am the cat burglar if you like. I get on my knees or back at your command. I am your fucking property. I have no choice but to mind you in almost all things. There are two realms that even the Master cannot control. You won't tell me what to think, and you are denied interfering with how I endure this bullshit brutality I call my shitty life."

Leo shook his head. "I am not trying to control you, Christian Axel. I am trying to help you. Why can you not see that? I ask you to stop this scouring because I love you. This has to stop, or I will lose you."

I felt the fury rising in my chest as the fires of hell broke out in my eyes. "Leo, you took my tools away. I have given you what you wanted. Now I want my sandpaper and razor back. They are for my preparation to play the sperm pocket for the nasty lust that I must attend. That is my fucking job. You would never take a carpenter's hammer, then wonder why the house is not built. Yet, you seem to think taking my things is a kindness. How the fuck do I function as a whore without them, Gott damn it. I want them back."

Leo grabbed my leash and jerked it hard. "You stop this right now, Christian Axel. You are getting yourself worked up. We drop this discussion, I mean it. You know better than to act like this. I want you to stand up for yourself, but not when it is over something you intend to do self-harm width. I will give you anything else you desire, but the sandpaper and razer, not happening."

I snapped at him. "You don't love me at all, Leo. You steal from me like all the others. Only when I pay them back with my flesh, do they at least give me the honor of returning what they stole. You can stop mocking me too. Bad enough you are trying to poison me with the pills without adding this parroting everything I say. I am getting tired of it. Where the hell did you learn your manners?"

Leo's eyes went wide in shock. "What is this you say? You think I am repeating you and slipping you poison? Oh hell, meine Heart, you are severely psychotic. Okay, just calm down, Christian Axel. No one is trying to hurt you. We are going back home now. You follow Leo in silence. Can you do that for me?"

I scoffed. "Then you give me back my things, thief? I can be quiet as death, but you won't shut the fuck up. I don't understand how you are not waking up half the Haus with that yelling you do." I crossed my arms.

Leo nodded. "Okay you come home with me without quarrel or noise, and I will think about returning the sandpaper."

I snorted. "And my razor. I want them both. They are mine, not yours."

He put up his hand demanding my silence. "The razor too. Now, I mean it, I want no trouble out of you."

I smiled with glee that this was finally settled. "I get my tools and you will have no arguments from Mad Maxx, Leo. Thank you for the mercy, and you don't have to shout. I am right here. I can hear you."

We returned to our apartment, and he locked all the bolts behind us. I knelt at his couch with a smile on my face. I could see the disease rising along my ankles, but I wasn't worried. Leo was going to give me back the tools I required for fixing this recurrent problem. I started to become a little upset when he went to his phone and called someone rather than fetching my property.

I waited with some agitation while he whispered to the person. I felt the room sway a bit, and the thinking of everyone downstairs was giving me a headache. It is quite bothersome to always hear the thoughts of the people in the Haus. They would get so loud sometimes; I couldn't listen to myself over them.

He hung up his phone and returned to stand over me. "Christian Axel, I asked you to stop kneeling in your own haus."

I shook my head "And I asked you to return my things to me, Leo. Where are they? I need a shower."

Leo frowned then sat down. "I kept your cure elsewhere. I called just now to get them sent to me. We will have to wait a bit."

I sneered. "I bet you the sonofabitch stole them. I will never see my things again. You shouldn't allow others to hold treasure, Leo. They always want their taste of it." I got up and began pacing, watching the door as I wrung my hands.

I was feeling anxious that this foul thing was going to get deep into me before I could put a stop to it. I looked at Mad Maxx and Max. They too were feeling nervous.

Mad Maxx frowned. "Do you hear that? It is that electrical sound again. Do you think they are working on a new roof?"

I scoffed. "Using power tools at night? Don't be stupid. Likely it is only rats in the attic you hear, brother."

Max shook his head. "Do rats use electrical outlets to run their drills? I think not. You both are speaking nonsense. I know what that sound is. That Cora is running her vacuum upstairs is all."

I giggled at that shit. "Really Max? Do you see the time? Not even that horse faced bitch would be up cleaning her Haus at this hour. Besides, she has the black collars run her appliances."

Mad Maxx nodded. "Mad Max is right. I bet she has that Olaf up their right now running everything she owns to make us nuts with the noise of dem."

I shot a look at the masochist. "News flash, whip boy, you don't need a vacuum running at midnight to be bat shit crazy. Fucker you are ripe for the squirrels winter stash you nut."

I stopped pacing and wringing my hands when Leo suddenly laughed out loud. I looked at him with curiosity wondering what was so funny. He immediately covered his mouth with his hand and pretended to be looking for his romance novel.

I shrugged. "I think people are strange, brothers. They should all be put into a zoo and examined."

Max smiled. "Isn't that what this place really is Mad Max? The monkeys wear jeans and the horse sits on the throne."

Leo snorted and chuckled loudly. "Okay, I must ask you to stop this conversation meine Liebling. You are killing me with humor. Do you speaking to yourself often?"

I looked around the apartment. "Who are you talking to Leo?"

He stopped laughing. "Christian Axel, there is only you and me here. I am asking you. This conversation with yourself is quite the comedy."

I backed up a bit. "Conversation with myself? Leo, do you have a fever? I am not saying anything. I am quietly waiting for my things."

Leo frowned then his expression went to worry. "Oh dear, this is worse than I ever imagined. How did I miss this? Meine Liebling, can you sit with your Leo for a bit? You need to calm your agitation. This pacing and speaking are only getting you more worked up."

I shook my head. "I will be peaceful when I get my shower. Is this man going to take till dawn? Where the fuck does he live, China? Is he having to walk down the Alps or something?" I pulled at my sleeves whining a bit.

Leo patted the seat next to him. "Please meine heart, sit down with me. Your cure is on the way. A few more minutes won't kill you."

I grumbled that it could land me in my grave, but I went and did as he asked. I couldn't shake my anxiety though. I rubbed my jeans with the palms of my hands pretending they were my lambs. I did my best to not lose my temper that this courier was the slow ass. He was wasting my time.

Within the next half hour there was a knock at the door. Finally, damn. Leo rushed to answer it. I sat on the couch trying to get a look at the bastard that was delivering my stolen items. The black collar was dressed strangely. He wore a black cowled robe, and I couldn't make out his face. I was a bit freaked out that this weird motherfucker had his filthy hands on my tools.

Leo thanked that nutball and took a paper sack from his hands. He then slammed and bolted his door shut, appearing a bit nervous. That made me even more afraid. What the hell was that shit all about.

I stood up wringing my hands. "Give me my things, Leo. I need the water. Enough of this. I can feel it digging into me." I stomped my feet trying to knock that nasty crap off me.

He walked over and put the bag on his coffee table. I watched as he took out a spray bottle with deep blue liquid in it, then a small mason jar filled with a strange orange colored substance. I stood there waiting for my sandpaper and razor. When Leo didn't produce them, I grew more than a little irritated.

"Where are my things, Leo? You said they were on the way," I shouted at him.

Leo picked up the bottle and jar, holding them out to me, "These are the cure you wanted. I called the ones that put that magic into the pool water. It cost me a lot of money, but they sold me a bit of their mysticism. This blue one is the same as in the pool. You spray it on yourself in the shower and it will kill the disease like it did downstairs. This other orange one you drink. If you consume all of it, then the disease cannot hook onto your flesh for a full day. Take them meine Liebling. It is my gift to you."

I took the items from his outstretched hands. I held the blue water to the light. It was no doubt some of that magic from the pool. I put it on the table while I opened the mason jar and sniffed it. I opened the spray bottle and took a whiff of it next. The chemical smell from the pool filled my nostrils.

I frowned. "This stuff smells like orange juice, Leo. The blue one is authentic, but the other is nothing but breakfast drink. I think these sorcerers ripped you off if you paid a lot for that citrus stuff."

Leo shook his head and crossed his arms. "Nein, I went right to the source, I assure you. Here let's try the magic water together. I could use a shower myself. You don't mind if I join you in your bathing do you, meine heart?" He smiled with a wink.

I looked at him with a startle. "Huh? Why would you do that? I can provide the bath service if you require such a thing."

He chuckled. "You can give me the bath service if you wish, but I will return the favor. Come on meine Liebling. It will be fun. I can help you spray on the cure and kill the foul disease, ja?"

I shrugged. "Okay. Seems like a lot of trouble for you, Leo, but if that is your pleasure then so be it." I picked up the blue cure bottle and headed for the spare bathroom to start the water for our shower.

I turned on the hot facet and watched the steam rolling. Leo came through the door already disrobed. He saw the steam barreling out of the tub and rushed forward turning it down. I frowned while watching him turn on the cold tap evening out the temp from scalding to tepid.

"Leo, damn it, now I have to wait longer. You wasted my time with this bullshit. The water must be hot to kill germs this tenacious." I started to reach for the handles to adjust the setting, but Leo grabbed my leash preventing me from reaching them.

"Nein, Christian Axel. You must have faith in the cure water. That pool downstairs is not boiling. I cannot stand in the hellish heat like you can, and I share this shower with you. Now strip down and get into the damned tub for this treatment of your ills. When you see it works then I will ask you to take the preventative they sent as well. If both works, I will make sure you always have these tools for your job." He took the water sprayer from my hands.

I glared at him suspiciously but took off my clothes. I wondered if he and these others were pulling a trick on me. The blue water smelled like the good stuff, but I wasn't convinced it was the same magic. Leo got into the tub and pulled me in with him.

Leo had to stand near the handles because I continued to whine the water wasn't hot enough. It did feel a little better, but I could see it wasn't doing shit to kill the monstrous filth climbing my legs. He tried to get me to focus on attending his cleansing service first. I did as he wanted but couldn't stop worrying the illness was gaining on me.

When I finished his scrubbing, he took the rag from me. I nearly fell out of the tub when he started to scrub me with the soap. I protested that a Master wasn't supposed to give the bath service to his submissive. He told me if he was the boss then he could do as he pleased. I endured his gentle scouring of my flesh with confusion. This Dominant was a weird cat.

After he rinsed me off, he grabbed the blue spray bottle and told me to tell him where I could see the filth. I pointed it out and he immediately doused the areas heavily. The smell of that chlorine filled the muggy air around us. I felt immediate relief. I looked down in astonishment as that nasty crap fell off and washed down the drain.

"How did it do that," I yelled out in excited shock.

Leo chuckled as he continued spraying me. "I told you; these fellows are the mystics. They know the cure for your sickness. It is right here. Anytime you feel unclean you spray the germs with it, and they die and wash away. You need not remove the flesh nor take the boiling showers anymore. There is no need for the razor or the sandpaper if you have this handy little blue bottle. You run out, let Leo know and I get you more. I will always make sure you have enough to keep you well and healthy meine Liebling." He kissed my forehead.

I smiled with glee. "Thank you, Leo. I apologize for thinking you were a thief. I have no excuse. I suppose I must beg your punishment for being the fool to doubt your love." I looked down in embarrassment at my impulsive judgement of my lover.

He nodded. "That is okay, but your punishment is to give me a kiss for this gift, and you must take the preventative the moment we finish this shower."

Ma eyes went wide. "No thudding? Leo, I called you foul names and questioned your manners. That is insolence of the worst kind."

Leo crossed his arms. "I told you I don't think it is insolence for standing up for yourself love. You thought I took something from you, and you called me out. I took your sandpaper and your razor too. I merely replace them with the right cure is all. Now for that kiss?" He leaned forward and I kissed him with much eagerness.

Oddly, the electric feeling from the pool filled my mind with vigor all over again. I began to paw at him with much wantonness. Leo tried to gently push me off him, but I continued to come at him, unable to stop myself from touching him everywhere.

He yelped in shock when I dropped to my knees and lustfully engaged him with my oral skills. It was as if something was possessing me. No matter how much Leo protested, though I noticed his cock didn't, I would not relent. My own manhood was fully alert and ready for adoration in return from my lover.

Leo pulled me to my feet unable to resist my aggressive behaviors any longer. He then engaged me in wild kissing and licking my mouth for a moment before dropping and dragging me to the tub bottom with him. He positioned me on my side, only breaking from our passionate kissing long enough to prepare me for his comfortable entry.

I gasped out in some mild pain as he took his mount. He paused his entry allowing for my adjustment. A true mercy none of the other many Dominants ever granted. In his place just under me he was able to keep his lips to my own. He kissed me with gentleness taking it slow waiting for my trembling to stop.

I gasped out again into his mouth when he reached around taking my manhood into his hand. He held up his own pleasure while he stroked me to a heavy pant. Then, as before in the pool, with a perfect rhythm he was able to engage in his thrust and grant me sexual thrills at the same time. I could barely breathe with the water pouring down on our heads, and his expert touch driving my lustful interest.

During this hedonistic intercourse he would lick or kiss my ear. I often turned my head to kiss him back with much vigor. Our recent lovemaking slowed the climax, for both of us, of this second much more eager act in his bathtub. I was stunned that for a change I wasn't praying he would reach his satisfaction quickly.

However, he did reach it faster than I did this time. Leo let go of me and rapidly increased his thrust unable to contain the buck of his reaching apex. He wailed out loudly in pure ecstasy as he orgasmed. I tolerated his finishing of his pleasure assuming that was just too bad for me.

Then I nearly fainted when Leo uncoupled quickly and pulled me to sitting. He went on his knees and dropped down between my legs. Without qualm or hesitation, he brought me to my own screaming climax with his talented mouth. I all but drowned in that water pouring into my panting mouth from the force of it.

My eyes had been forced shut during this most amazing sexual release. When I opened them Leo's face was above my own. He kissed my lips then pulled back with an evil grin.

"You are the beast, meine Liebling. I will not be able to keep up with you at this rate." He kissed my neck.

I laughed out loud. "I apologize Leo. I don't know what the hell got into me."

He chuckled. "Well, I would say I did but this time I think we got into each other."

We both howled at that while he reached over and grabbed the rag and soap. "Are you done seducing this old fool with your charms? If so then we can clean up and use the cure right this minute. If not then, I will hold off, and call the paramedics too. You are going to put me in traction, meine Hase."

I nodded. "I am done, I think. You look awful good to me still." I gave him a coy look for a tease.

Leo covered his mouth dramatically. "Shit, I just went on a diet. Oh well, looks like I will have these love handles for all my life. I have too much of a sweet tooth for my Christian Axel to deny such a delectable treat."

I chuckled. "I have been told there is only twenty-five calories in the orgasm, Leo. I think your safe from getting fat from blow jobs." *Good thing Meine Liebe or I would have to be hauled around by a crane as many as I have been given. Yikes and yuck, ja? I nodded because I was thinking the very same thing about myself. Of course, half the time that was the only calories I got in Debbie's house. Damn.*

He roared in laughter at that. "I don't know what bothers me more, that you know that fact or that I am relieved to know I have a lite lover."

Leo and I finished our cleansing with much teasing and funning each other. He sprayed my flesh with the magic water. He said he wanted to be assured nothing had been missed in our first pass with it. I got out and dried off then attended to Leo. He told me there was no need to re-dress him. I watched him leave the bathroom and I quickly put on a pair of breeches and a blouse, but I didn't bother to button it, nor to put on my boots.

I hid the blue bottle cure in my bag. I rushed to the living room to be near Leo the second I got the room cleaned up from our follies. I found him sitting on the couch reading his romance novel in a pair of boxer shorts and fuzzy Frau house shoes. He looked silly which made me giggle as I came in to join him.

He looked up from his book with an expression of fake indignation. "What is so damned funny about a man being comfortable in his own home? You come here and sit with me. You drink the preventative, and I will read you this chapter from my novel. This woman in the story is quite the lover. You should hear of her exploits, Christian Axel."

I joined him on his sofa shaking my head. "Why is she so seductive? Is it a story of the Female Priceless trying to break her collar?" He had my attention if this were the case.

I was thinking if there was a manual on another like my Frau would be, I needed to study it right away. Leo stared at me for a moment, he seemed confused, then pointed at that orange juice smelling stuff on his table.

"Ja, I guess she is though I never realized it. Drink that and listen for a bit. You may find her techniques of interest." He watched me pick up the jar and start swallowing it down. It tasted the way it smelled, like orange juice.

I wrinkled my nose upon the first gulp of it. "Yuck, I hate this citrus drink. I guess like any medicine it was bound to be foul to the tongue. I wonder why all the cures are so nasty." I investigated the jar to see if I could spot any mystical signs in the fluid.

Leo shrugged. "Well, I told you meine Spatz, everything has a price. The cure is battling a tough foe.

War isn't something that should taste sweet on the palette. I should know, I have been through a couple of them. Not good times. This last one tore the country apart, ja? It's aftermath will be felt in her people for many years to come, I think. something so traumatic cannot be solved in a day. The Motherland is a lot like you, meine hase. I imagine you will be drinking lots of this preventative and going through many blue bottles before you are completely whole again." He reached out and ran his fingers through my hair and I noticed his arthritic hand was purple with bruises.

I narrowed my eyes at that. "I see that arthritis is getting worse in your knuckles Leo. Did you maybe hit it on something? Like a door jamb or a jaw maybe?"

He looked down at his injury. "Oh, this. It is nothing to worry you about meine heart. Now finish that up and I will read." He opened his book, changing the subject.

I listened to his tale of a woman juggling many lovers. I wondered to myself about the identity of the one Leo had nearly broken his hand on. He wasn't fooling me. I could clearly see he had punched someone with much force with that bruised hand of his.

The dishonorable Julius came to my mind. I thought maybe Leo had caught that man out and thrashed him for his bad manners. That made me smile, but I knew that was not all that bastard had coming. Leo got his revenge. It was Mad Maxx's turn to show that motherfucker what happens to those that rape me without rights to that privilege.

I finished the orange drink while listening to Leo read. Within only fifteen minutes I felt the boy get heavy with fatigue. I leaned back into the sofa feeling that my eyelids weighed a ton. Leo took a pause from his reading when he saw my head drop from my nodding off a moment.

The sudden falling sensation caused me to startle and yelp out. I was confused and upset for a few minutes not recalling where I was located. He calmed me by reminding me that I was home. He put down his book and pulled me to my back, laying my head on his lap. I didn't fight him. I couldn't. I was too tired to even move. I guess the excitement of the day was taking its toll. I wouldn't have been able to rumble with a mouse.

I closed my eyes drifting off to a deep slumber. Leo was playing with my hair and speaking gentle words of love just as the void came and took me for her own.

That night I had the oddest dreams. In one, I turned into a bird and flew down Leo's hallway. I landed on his bed. Leo was singing to me. Then he said I was the prettiest Spatz (sparrow) he ever saw. He stroked my feathered breast. He kept repeating he wanted me to sleep peacefully and let the magic juice calm my fears.

In another, he pulled my head up from a comfortable pillow. I wanted to talk to him, but my tongue was stiff with fatigue. He held a cup to my lips and demanded I drink in the cure. I could taste the oranges as the citrus stung my inner cheeks.

I tried to spit this foul stuff out, but he pinched my nose and covered my mouth until I swallowed. I was not sure why he was being so cruel. I decided this was not the Leo I knew. This had to be a version my sleeping brain made up. I gave no further quarrel once I determined this was an illusion.

When the sunrays from his window touched my eyelids, I awoke with a start. I looked at the clock and saw with horror I had overslept. My lambs had missed their breakfast. I turned to awaken Leo to take me right away to fix this terrible mistake.

His side was empty. I panicked and jumped from the bed running all over the apartment. Leo was nowhere to be found. I was left alone, and my lambs were suffering from hunger.

I rushed back to our room and hurriedly put on my boots. I didn't bother to change my clothing nor button up my blouse as I ran back for the kitchenette. I dug around till I found the lock picking tools for his door. I needed to find Leo. I assumed he was in the Great Hall eating his breakfast.

I broke out of the apartment in less than ten minutes. I dropped the screwdriver and hairpin onto his floor without care. I took off rushing down the hallway at full speed.

I didn't bother with my sunglasses, nor even a coat. The fact that I was barely dressed never once crossed my mind. All I could think was of my babies calling and their suffering.

The silver collars dropped to kneeling as I passed. I jumped over, skirted past or even pushed them out of my way. My panic was building to painful levels. I made it to the Great Hall quickly despite the many submissive and trainee roadblocks.

I looked into the huge room from the entry. The hall was packed. I scanned the tables but could not spot Leo in any of the Dominant dining groups. Some of them saw me standing there craning my neck with wild eyes looking them over. They whispered and pointed. A few laughed at my obvious anxiety.

The attending black collar saw me and rapidly asked me to leave. Without their Dominant no silver nor black that don't work the Hall are allowed to be even at the doorway.

I asked him if my Master Leo was at his breakfast. He shook his head and said Leo had not been to breakfast for the last many days. I thanked him for the information then left without quarrel.

I wandered down the hallways trying to find him among the crowds. I was becoming increasingly upset. I was beginning to think he had abandoned me or worse whoever he struck had come seeking revenge. My step quickened with each passing moment of my not discovering his location. The more upset I got the louder the residents around me seemed to get.

I covered my ears trying to drown out all the noise they were making with their constant chattering. A silver male saw me moving toward him rapidly holding my head like that. He grew nervous and dropped to a kneel seconds before I was on him. I tripped over his flesh near falling face first to the floor.

I banged into one of the Dominant's apartment door and grabbed the short entry wall in an attempt to prevent my fall.

The Mistress that lived there thought someone was knocking. An unattractive, middle aged brunette FemDom opened it to greet her company. I stood there in shock unable to find the proper words of apology for disturbing her.

She smiled with thrill. "Mad Maxx, Ah, what a surprise. Louis, come quick. I cannot believe it. This must be the repayment of that favor you did for Claus. It is worth more than what he asked of you. He has sent a rare gift. The Priceless is here." She reached out and grabbed my chain leash before I could flee in terror.

I pulled back with strength trembling. "Forgive me, Mistress. This was not a planned visit. I beg your mercy for disturbing you."

The Mistress ignored me. "Damn it, Louis. Hurry, he is trying to get away. The rumors are correct. He is a fighter. How exciting. Come help me get this collar inside. I will call Dominic and Claudia. I want to show off our gift leash. You were right. Claus is indeed generous. I cannot believe it. We get the Priceless for a whole night. Ah, what a happy day."

A short heavy-set man with black hair came rushing into the room with his eyes wide in surprise. "Nelda, did I just hear you say, oh Meine Gott. Glory be, it is the Priceless. I thought you were funning me Nelda. Fuck, get him inside before someone else snatches him. I cannot believe our luck. I am not going to call Dominic and Claudia my love. They will only desire to take a turn with something that belongs to us exclusively. We call them in the morning, before we return Claus his property, after we have had our fill. We only have twelve hours to get our taste of the forbidden metal. The showing off can wait." He came running at me and his Frau FemDom with a huge smile on his face.

I let out a wail. I was frightened near out of my skull of being raped by these misunderstanding Dominants, and maybe their friends too. My sudden loud screaming caused Mistress Nelda to startle. She loosened her grip on my leash for a tiny moment, appearing unable to understand my horrified behavior.

I didn't miss my chance to get the fuck out of their with my breeches on. I pulled with all I had and broke free of her hold. I turned and ran full speed jumping every kneeling silver as if they were hurdles at the Olympics.

I could hear Master Louis and Mistress Nelda yelling in chase behind me. Master Louis kept demanding that I "halt at once."

Thankfully, I was much faster than those middle-aged Dominants. They were in no shape to win a race against the lithe runner Mad Maxx.

I managed to lose the lusty Dominant and his status seeking FemDom by hiding in the darkened dungeon stairwell. I watched the two of them run past the doorway without seeing me huddling there.

I was waiting for the coast to clear when I heard two familiar voices approaching the stairs. I peered through the darkness of the steps to see Peter and Gretta fast approaching my hiding spot. I couldn't believe my bad luck.

I decided to take my chances that Master Louis and Mistress Nelda had given up their chase. I took off up the top few rock steps.

My hope was that I had not been spotted by my father and the Head of the Voting Council. Peter shouted out at me just as I hit the doorway shattering all my desires to avoid this confrontation.

"Halt, stop right there you little bastard. I want to speak with you Maximillian. You better kneel this fucking minute," Peter shouted out with much anger from behind me.

I ignored his demands. With all the speed I had left I fled the Dungeons. I saw Peter, like Master Louis before him, take off running to give chase. This morning was turning into a real nightmare.

I ran down the main hallway without pausing. I was headed for the front stairs ready to get home and await Leo's return. I simply couldn't outrun the whole fucking Haus. My lambs would just have to forgive me. The way I saw it, if I got myself ripped to pieces by greedy Dominants more than just my little ones breakfast would be late.

I was almost to the stairwell when I saw Vilber and Olaf standing at the doorway as usual. I raced up to them refusing to look behind me to see if Peter was on my tail.

The black collar brutes stared at me with expressions of disgust as I approached them near out of breath from my pursuers.

"Look here, Vilber. I think Mad Maxx comes to tempt me. See told you. He wants to let me finish fucking him at last. I knew he couldn't stay away from his dream lover Olaf. He is half undressed and already sweating from his dreams of me inside him." The black collar guard laughed with much evilness in his tone.

I narrowed my eyes and frowned. "I think you forgot already what happened the last time you tried to fuck me Olaf. I have no time to get into this banter with you, fool. Have either of you seen Master Leo or know where he is right now?"

Olaf shot a look of humor at Vilber. "And what if we do?"

I cast a nervous look over my shoulder to see that Peter was coming my way. "Shit, I don't have time for games Olaf. Where is Master Leo?"

Vilber chuckled with evil joining his partner with the hateful comments. "If Olaf tells you, what will you give him for the information? I seem to recall you are an equal service for equal service submissive. Nothing is free with you, then it will not be with us either."

I shot Vilber a look of hate. "I tell you what I give you if you don't tell me, you motherfucker. I will give you a one-way ticket to the fucking river Styx. Now, where is my Master, you rat bastards."

Vilber smiled at Olaf. "You know what we should do, brother. Get rid of this rodent like we did to that fucker Abelard, ja? Even the demon Mad Maxx cannot outfox the bullets."

Olaf grinned with wickedness. "Hell, ja. You get the door."

Before I could run, Olaf rushed forward and grabbed my leash. I struggled and yelled while punching the brute with all my strength.

I watched in pure shock as Vilber opened the door of the Haus. Olaf pulled my leash with force, sending me face first to the floor.

I tried to get up, but the brutes grabbed my upper arms. Without hesitation they tossed me out the open door, then slammed it shut behind me.

I rolled helplessly across the entry porch right onto the front lawn. As I came to rest at last, I screamed in terror. I was outside the Haus without a Dominant. The guard was about to shoot me for trying to escape.

I got up with fear driving my worn muscles. I rushed to the door trying to get back inside. The fuckers had locked me out. I pounded and kicked the door wailing out near mad with panic. Within moments I heard a loud popping noise. The Guard are shooting at me.

Terror blinded my better senses. They never would shoot me through the damned door. All I could think was, I had to get out of there.

Like a fool, I took off running for the parking lot. Bullets struck the ground around my feet and whizzed above my head. I screamed for help with each sound of the rifle. I had never been so scared in all my days. This was surely the end of Mad Maxx.

It was at that moment I realized the shells had missed earlier in my run by miles. Suddenly, the shots were getting closer to making their target. I wailed and turned tail running back the way I had come. I understood at last I had been running toward the shooter all that time.

The sniper's weapon began to miss me widely once more. I kept running right past the front of the Haus. I decided to try to make it to the playground. I thought I could maybe hide behind a tree from this shooter hell bent to send me to join Abelard.

As I approached my destination, I felt my knees wobbling. I couldn't catch my breath. I feared I was going to fall, then be unable to run any further. The bullets still hit the ground around me, and the sound of the rifle continued but it was now in the far distance.

I collapsed right at the foot of the swing set helpless to run anymore. My bid for life was over. I could go no further without rest. I rolled up into a ball and waited for the blow of the bullet that would send me to my grave.

My heartbeat in my ears, and the sound of my panting was loud enough to drown out any other noise. I laid there for what seemed like hours trembling in tears at this most unexpected end to my struggles. It took quite a bit of time for me to notice that the sniper had stopped shooting at me.

I didn't move for another several moments even after this discovery. I thought maybe he was waiting to get a clean shot to my head.

When the birds began to sing their songs around me once more, I decided I couldn't lay there forever. I lifted up with much slowness, ready to dive back at every sound from the breeze to the sounds of the distant sheep calling in the fields.

I looked around and didn't see any Russian brute standing around waiting to blow my brains out. I stayed sitting and pulled my knees to my chest unsure what to do.

I couldn't get back inside the Haus without Dominant protection. I couldn't dare go anywhere else in case the shooting would start again. This time maybe I would not get so lucky, ja?

There was a deep patch of woods at the back of the playground. I wondered if I could hide in them until I spied a Dominant coming by. I thought maybe if I politely asked, they may take pity by leading me back to Leo.

The wind whipped at my sweating flesh sending chills down my spine. It was autumn and I had not properly dressed for a harsh German season.

I shivered and my teeth chattered. I looked down and began to button up my blouse. It was silly to think that would help much but having it open wasn't helping either.

"Leave it like that. I think it looks better that way," a voice called out startling me from my buttoning task.

I looked up and felt my blood freeze in my veins. It was Julius standing there smiling and looking me over with a cruel smile on his face. I swiveled my head all around and saw that he and I were alone. No one was around to aid me or hear me scream.

I shot him a look of fear. "Can you take me back to my Master please? I thank you for the mercy of it." I did my best to quell my rising terror at my compromised position with this huge, fit man.

Julius nodded, still smiling with evil. "Sure, I can, Mad Maxx. I will be happy to hold your leash. First though I would like to visit with you a bit."

I didn't like the way he said that. "I thank you for offering such an honor to one so unworthy, but I must beg that you return me immediately. My Master will be worried if I am late."

He chuckled. "Will he? I bet he is more than a little worried, my love. He doesn't even know where you are, does he? Especially since he never thought you would be outside the Haus without him. Awfully nasty of Olaf and Vilber to throw you out like they did, ja? You are lucky those bullets missed you. Amazing, all the shit you have managed to survive. You keep hanging on to life no matter what the Dominants of this Haus throw at you. Well, looks like your luck has finally run out Mad Maxx."

I shivered from the cold and his cryptic words. "You saw Olaf and Vilber's trick? Would you tell the Elders what you saw?"

He nodded. "Ah, ja, I did see. I have been following you quite a while now. There are not many places you go or things you do I don't witness. I will not be telling the Elders about the black collars any more than I will alert them to your killing Stefan."

I nearly pissed myself in pure terror. "I apologize, but I don't understand. I know nothing about Master Stefan. Has he passed? There is no rumor of such a thing anywhere I have heard. Wait, if you think I did such a foul thing why wouldn't you tell the Elders."

Julius looked at a passing bird in the sky. "There is no thinking Mad Maxx. I happen to know you killed him.

Clever boy, you hid the evidence very well. In another life I could actually respect and even call you a brother. Sadly, in this one, you are nothing but a game piece. I must say a fine piece all the same. I have enjoyed you a great deal. A real fighter and the sex has been the best I have ever had."

I looked at the ground with my ears burning with anger. "I knew it. You are the rapist. You are the one that has been following me. Why are you doing this to me, Master Julius?"

He laughed loudly until he coughed from the force. "Ah, it was nothing personal my boy. Though it was a real pleasure without a doubt. That fucking Leo has this coming. All the Elders do. They hurt my Beatrice. She is stuck far away from her family and friends thanks to all them ganging up on the sweet girl. Over a fucking bunch of silvers too. What a joke. Your kind are nothing but beasts for our lusts. I was going to just kill the Elders one by one till I saw them send their Priceless collar out to kill Beatrice's cousin Stefan. Then I realized they were on to me. Well, I heard their message. They took everything from my girl, so I take all that matters to them. A little pepper spray and a little spit and I was able to enjoy a rare treat. I have thrilled at this revenge more than you know."

I winced at that. "I think I am more than aware how much you have. So, you come to kill me this time, correct?"

He nodded. "Ja, my little dove. It saddens me to have to give up my plaything so soon. We had only just gotten to know each other. Oh well, your death will hurt the Elders deeply. Then once they are done with their moments of grief, I will pick them off. Starting with that tattle tale Leo. Now, come with me, my boy. I want to taste you one more time before sending you to oblivion. You can try to run but they will shoot you. If you fight, then I will just beat you till you stop. There is no choice. Come quickly and I promise to make your death swift and painless as possible." He held out his hand with a grin.

I sniffed back my tears, realizing he was right, there was nowhere to run. "You cannot just rape and murder me here. Someone will see you. Let me go and I will never breath a word of what you have done to me or of your plans." I knew he wouldn't buy that, but I had to try.

Julius giggled. "I don't intend to fuck you where anyone can come by and take photos, Mad Maxx. You come with me, and I will be gentle this time. No need to bust you up nor blind you this time. I think you should consider I offer a merciful end to a most brutal existence. You dare to deny me, and I make it end the way you have lived it. Come with me now. I tire of this discussion and sooner or later someone will see us." He smiled as I stood up, keeping my gaze to the ground.

"I give you no quarrel, Master Julius. Do what you must." I trembled slightly as he came forward and grabbed my leash.

He patted my head as if I were a hund. "Good boy. Now follow me in silence or it will not go well for you."

He took off toward the patch of trees with a gleeful smile on his face and me hostage pulled behind him by my leash.

I was forced to walk briskly or fall to be dragged. Master Julius had long legs and was in a rush. I shot looks all around hoping against hope someone would see this horror and save me.

To my dismay he continued to lead me deeper into this tree-filled area. No one was anywhere around to notice a damned thing amiss.

The sun was soon blotted out by the heavy forest cover. Dark shadows were everywhere. I shivered with dread. I had been correct in the identity of the rapist. I had figured it out too late to end him before he came to end me.

I could do nothing but stare in terrified silence at Max and Mad Maxx. This was as bad as it could get. I wouldn't get the mercy of my death without the dishonor of one final violation by a man. I shook my head full of despair at this most unappreciated indignity.

Master Julius let out a gasp of thrill "Ah, I thought we were lost for a moment there, Mad Maxx, but here it is just as I remembered. Beatrice's used to meet me here years ago when we were young Dominants, and I came to visit her. The wishing well." He stopped in front of a dilapidated rock water well long since abandoned with the invention of modern plumbing.

I dropped my gaze to the ground noticing many of the stones from this ancient drinking fountain had broken loose and rolled across the forest floor. Master Julius pulled me harshly to follow as he walked over and looked down into the deep hole.

"She and I used this well to dispose of many a young silver collar. Ah, those were the days. Back then, when I would come visit her, she would bring the prettiest, freshest male or females to taste for their first time with me. Oh, how they would cry and scream for mercy as we soiled them in every way. When at last there we tired of looking at them, we would break their little necks and dump them in. I wonder, with all the common collars down there, will they appreciate being joined by metal of the finest quality? I doubt they will care. Priceless or simple you are all just a bunch of whores put here to serve your betters. None of you can do a thing but bow and scrap. Cattle have more will. Now, Mad Maxx when I finish having my fun with you, I will throw you down this well. You get to join your unworthy brothers and sisters in a watery grave. Don't you give me any trouble, or I will not bother to break your neck first. I heard a rumor you cannot swim though you do enjoy your Master when in the pool, ja?" He turned around to enjoy my look of terror.

I trembled feeling sick to my stomach that he had seen that business with Leo the night before. "As you wish Master Julius. Is their nothing I can do to change my fate?"

He pursed his lips then smiled with lust in his eyes. "Tell you what Mad Maxx, you show me the best of your fabled artistic skills then maybe I consider chaining you up here for a few days. I will kill you in time, but you can buy a little more time on earth this way. We have a deal?"

I nodded knowing damned well he was lying through his teeth. The second he reached his orgasm he was going to send me to bottom of that rock well. I didn't move as he undid his pants demanding I drop to my knees to get him ready by my mouth for his couple.

I knelt without quarrel as he approached. I kept myself calm as I began to employ only the finest of my skills to his manhood.

Master Julius initially stood there saying many foul sexual things but, in moments, he was overtaken by his lustful drives from my skills. He held my chain leash tightly in his hand but closed his eyes, tilting his head to the sky moaning in delight.

I kept my eyes open and to the ground while suffering this indignity. I very carefully reached out and grabbed a large rock that had fallen from the well wall. I pulled it slowly next to the boy, being careful not to make a sound or break from my task of keeping Master Julius blissfully unaware of my activity. Well, the other activity of grabbing that weapon.

When he thought he was ready for his mount he commanded me to stop the blow job. He smiled and disgustingly told me he was thinking of keeping my collar for himself if he could be assured such "pleasures" for all time.

I nodded. "As you wish. Shall I get like a dog, Master," I said without inflection nor emotion.

He laughed. "Ja, I do enjoy the taste of your metal and this perfect service you provide. I can see why the Elders and Peter fight so hard over your charms. Get on your face Mad Maxx. Let me show you how a real man fucks."

I nodded and took the position of submission appearing subdued and compliant. He dropped to his knees behind me. I heard him spit then grabbed my waist trying to pull down my breeches.

I took a deep breath then without another second hesitation I picked up the rock. I spun around and bashed him on the side of his head.

His look of lust melted into one of shock and agony. blood splattered everywhere from the blow. I stood up quickly and hit him with the rock in the top of his head before he had time to wail out from the first attack.

Master Julius fell to his face dropping my leash wailing in pain. I stood over him watching him claw at his face and try to cover his head. He was blinded by the blood pouring into his eyes.

I laughed at the sounds of the demons rising within. "Nibble nibble like the maus. Who is that nibbling at my haus," I sang out loudly to him.

Suddenly Christian came out of the shadows. He jumped into the boy and threw all three of us Max shards into the wall. He looked at us trembling in terror at his hijacking the flesh against Der Hund's directives.

"You pussies watch the Master Christian teach this pig some manners. Equal service for equal service boys. Hold on to your dicks. I am gonna ride this wave of crimson right to the fucking ocean of nothing." He reared back the boy's arm and began to repeatedly smack Master Julius in the face and head with that rock.

We couldn't tear our horrified eyes from the slaughter scene. Christian pounded Master Julius's head to a bloody pulp. He was so disfigured by the force of his released rage that identification would have been difficult. Even the man's mother would not know her son.

His eyes bugged from the busted eye orbits and his swollen tongue protruded from his mangled jaw. It was beyond gruesome, Meine Liebe. To this day, I still have nightmares of the sight of what a horror Christian made of Master Julius's head with that rock. Blood and his brains were on everything.

The truest terror of it all is that despite the condition of his skull, the man was still breathing. Christian dropped the rock then dragged his limp frame with some difficulty across the ground. He took him right to the edge of the well. Without care that the Dominant was making whimpering noises, he pulled him up until he was leaning over the edge of the hole.

He giggled with the sounds of hell in his tone. "You go to join the poor innocent ones you thought nothing of, you sonofabitch. Maybe they will show you more mercy than the Priceless, but I doubt it. You of all people should have known better than to fuck with us. You were there when we sent the Head of the Haus to hell. Didn't you believe how deadly a couple with us was when you were a witness to our powers? Oh well, you can't look at anything with those eyes of yours anymore now that I have removed them. Maybe now that you are blind, you can finally see? Nein? Of course not. Your kind never learns. You can tell the devil and Xavier that Christian sends his love. Buh-bye, Master Julius." With that he pushed him into the well.

A few seconds passed then we heard the faint sound of a splash in the water at the bottom of the well. Christian went over and grabbed the rock he disfigured Julius with and dropped it after he finished tossing the Dominant.

He howled in glee. "Ah, look there Mad Max. The legend of the wishing well is true. I threw in a worthless thing and another worthless Dominant is suddenly gone from this earth. My wish come to pass, ja?"

Der Hund came from the shadows with his eyes on fire. "Christian, out of the flesh this minute. Your task is complete. Hand the wheel back to Mad Max. You mind me or be shattered."

Christian turned to look at the three of us. "Got to go schwuler boys. Catch you later. Enjoy cleaning up my mess. Like the frau's you all are, the man does all the work, and the women maintain the chores." He jumped from the boy and fled back into the shadows.

Der Hund turned his attentions to the boy. "Christian is not to be taken seriously. However, he is correct about one thing. This mess must be hidden. Mad Max, you get rid of all this evidence including the shirt the flesh wears. There must be nothing left of this murder scene."

I trembled from the deepening cold and the fear rising within. "Der Hund, we are in much trouble. I cannot get back into the fucking Haus with the boy. Those snipers are out their waiting to shoot us."

He frowned. "You are a fool, Mad Max. There is only the one sniper. If there had been many the boy would already be full of holes. You clean this blood and foulness up, throw it down the well. Then you walk a straight line from the playground to the Haus and enter the side door at the torture rooms. This sniper is guarding the front of the Haus only. There is likely one in the back too. Move quickly so no one sets off the dogs. I think however, they only use them at night. Now hurry the fuck up. The flesh will freeze out here, and Master Leo will worry if he returns to find the boy gone." With that Der Hund returned to the dark spots among the trees.

I rushed to the gory spot on the leaf strew ground. I picked up all of it in handfuls dumping it into the well. Then I took off my soiled blouse. I spit on it for several minutes until it was soaked.

I wiped up the blood best I could from the boys flesh with the bloody thing. I thanked Gott my breeches were black, unlike that fucking blouse which was white. No one could see the blood on them, or I would have had to run back to the Haus wearing only my boots.

When I was satisfied no evidence of a struggle or murder was visible, I threw the blouse in and took off back the way Julius had brought me.

I walked along holding my chest in a hug shivering from the cold but feeling better that I no longer had to be afraid of being raped and eventually killed by Julius.

I couldn't stop thinking of the man's gory face when Christian bashed it into pudding. I distracted myself with thoughts of my lambs, Der Makellos and my Leo. That made me feel peaceful and calm.

I approached the end of the forested covering at last. I stood at the edge of it feeling terrified. I prayed Der Hund was right on his belief there was only the one sniper. If there was another, the second I step into the clearing, my latest win in my struggles would have been for nothing.

I took a deep breath and braced myself for a mad dash to the side of the Haus. I was about to take my first step when I heard the snap of a stick breaking behind me. I turned around with a startle terrified Julius had escaped the well.

Pepper flew into my eyes and mouth sending me straight to my knees clawing at my burning face. I couldn't breathe or scream out. I was immediately helpless as the chemical caused my mucous membranes to swell.

My tears flowed like rivers when I felt the familiar hand of the rapist grip my naked upper arm. He dragged me back into the seclusion of the forest kicking and coughing.

I couldn't believe Julius had managed to survive. This was insane thinking, but I couldn't understand how else this horror could be happening if his corpse was floating in the well.

The unknown assailant stopped hauling me and dropped down tearing at my breeches. I sat up with all I had and grabbed at his face expecting to feel the mangled flesh of Julius's disfigured head.

Instead, my claws found the attackers face clean shaven with short hair just as before in Barnim's apartment. I was stunned. This was not Julius. The man backhanded my face harshly, sending me half dumb to my back in confusion.

I laid their unable to fight with any vigor, thanks to the blow and that fucking pepper spray, while the man got my beeches removed. He crawled between my legs, and I heard him spitting.

With all my strength, I attempted to struggle again even though I knew it was hopeless. I coughed to near a faint as I tried to push him off me. The strong, fit man subdued my weak flailing arms without much effort. He held them to my chest, pinning the boy to the ground with his heavy weight.

I gasped in agony when he brutally penetrated me with no mercy of any kind granted. I openly wept and whimpered as he began his harsh, rapid thrusting.

The nightmares in this place never seemed to end. I had already escaped one other attempted assault, barely survived a second half attack, only to end up the victim of a third horrific one. All within less than a few hours.

As I helplessly endured his brutal intercourse, I suddenly recalled Julius had not said a thing about Barnim's apartment when he spoke of his attacks spots. It was now painfully clear to me there had been two rapists all along, not just the one.

Apparently, the two rapists were working together, and I just killed this guy's partner. Surely, he saw me send Julius to his grave in the wishing well only a short distance away.

I assumed this man would finish what the dead Dominant had started after the second he had gotten his fill of his taste of my metal.

Chapter 49: Casper the Friendly Bank Robbing Ghost

The rapist finished his intercourse with a loud moan. I coughed and shivered from the cold and the pain of his rough couple. He fell forward onto me still pinning my arms with his hands. I laid their feeling fury within as he panted into my ear from his spent lust. I turned my head and bit whatever the hell part of his head I could get into my mouth.

He let out a loud groan and reached up, grabbing my throat. The bastard squeezed it till I released his flesh. I assumed by the feeling of the shape I had nearly bitten his right ear off in my attempt to fight back. The second I gasped out he let go of my neck. He lifted me and backhanded me with harshness but with not all the strength he was capable of.

The cold had nearly frozen my flesh. His blow stung almost as badly as that pepper spray did. He held my arms tightly and moved about but to my dismay didn't uncouple with me. I kicked useless to make any contact on him still between them. I heard him growl like a bear then I felt a painful stick in my right hip.

I spasmed and bucked realizing he just stuck a needle into me. He held me tighter trying to hold me still. I whimpered when I noticed him taking much time to inject whatever the hell was in that syringe.

Within moments, I felt my heart start racing, sweat poured and the panic rose within. I was sure, like Drexel's medications, I was about to be paralyzed. I thought he wanted to end my battling him so he could easily drag me to the vishing well. This seemed like the end of Mad Maxx for truth this time.

Then suddenly, all the pain in the boy went away. I assumed the needle had delivered a killing poison to the flesh. I couldn't see but I didn't care. It seemed nothing mattered anymore. I giggled and moved my arms that seemed to be light as feathers. I felt the weight on my chest lifting. Then the sky opened, and it began to rain. That didn't bother me either. I like the water it always makes everything better.

As the refreshing liquid poured into my eyes the burning was fading fast. All the things that had been worrying me were forgotten. I decided I was very happy about this death business. It was better than I thought it would be. I wondered why I had fought so hard to avoid it. No more troubles, no more stress, this was great.

I reached up and wiped my face, spitting the water out of my mouth. I found it odd that I appeared to need to breathe even in this state of being deceased. It was a bit of a surprise that being a ghost would require respiration. I never would have guessed that strange fact. The idea of my spirit needing the air made me laugh a great deal.

I sat up watching another wispy soul grab my breeches. I tried to figure out why this fellow ghost had a hood, with large sunglasses, on. The day was overcast without a strong sun. This guy must have really sensitive eyes.

I noticed he also wore a kerchief across his mouth and nose. I chuckled at that thinking the man must have been a bank robber in his life. I smiled at him with much humor as he approached handing me my breeches.

"Did they shoot you for robbing the banks? I bet the gunshot wounds hurt bad when you died, ja?" I asked him, trying not to giggle at how silly he looked in that getup of his.

He nodded then pointed at the breeches in my hands. I looked down at them with a startle unsure what the hell I was supposed to do with dem.

"Can we go see the lambs? I want to kiss them goodbye before we go to, wait, where do the dead go?" I looked around at the weird landscape of this place of the dead.

The Haus was there, but it was fuzzy. It faded in and out with many colorful lights blocking my view of it. The ground seemed soft. Nothing was familiar but I recognized it at the same time. I felt the nagging inside that I was supposed to recall something, but I was so confused. I giggled at the idea that the dead could have anything to remember. Life was over for Mad Maxx. I was kind of sad I never broke my collar, but then again, I never really got much of what I desired when I lived. Sucked to be me.

I looked back at my fellow corpse. "Do you know if we can fly or swim? Can we go through the walls?"

The bank robber spirit pointed at the breeches in my hands again. I shook my head. I saw no need for clothing. No one gives a shit if you are naked when you are dead.

He came forward and took them from me. I raised my eyebrow as he began forcing the pants up my legs. I giggled loudly when I understood he was trying to put them on.

"You are the prude. Can I try your sunglasses? I have a pair that Leo loaned me. I wish I had them right now to show you. They are very nice. Where did you get the ones you wear? Leo said he got his from Paris." I rambled out as I watched this strange ghost force my beeches to my knees.

He didn't answer my questions. I was confused when he came around the back of me and lifted me to my feet. I stood there feeling the world rocking and swaying as he pulled them all the way to my waist. Then he reached around and buttoned them closed.

I staggered and then fell back to the ground on my bottom. I laughed hard at that. I supposed it would take a moment to learn how to float around like that other fellow did. Being a ghost was more work than I could have guessed.

I saw the robber get my boots. He came and knelt, forcing them on my bare feet. I snorted and giggled at his struggling hard to get them on.

"You are putting them on the wrong feet. How come you are not speaking? Did they blow out your tongue when you robbed that bank? Is that why you hide your face? It is gruesome, ja?" I suddenly shivered with a memory. There was another spirit lurking around this place. One I sent to his grave.

I spun my head around in a mild fear looking out for the spirit of Julius. I was certain he would not be glad to see me. The burglar man nodded at my questions as he exchanged my boot to the proper foot.

I winced at this disappointing news. I was unhappy that even in death I was stuck without any reasonable conversations. The only ghost around had no fucking mouth. Should have guessed that with my luck.

The robber spirit stood up and took my chain leash. He hand signaled for me to stand back up. I nodded then attempted to mind him. I fell forward and then staggered around unable to get my legs to do as I wanted them to do. My fellow ghost came forward and caught me by my upper arm before I fell to the ground as I did before.

I smiled at him. "Thank you for the help, uhm, oh this will not work. I have no name to call you. I think that Casper will have to do since you have no mouth. Thank you, Casper. Did you say we can go see the lambs? I really need a nap before we go though. I am sleepy. Do ghosts sleep much? I hope not. I don't like all the nightmares, ja?" I held on to his arm to steady my wobbling legs.

He hugged me to his chest and helped me walk through the forest. I tripped many times, but my friend Casper was patient with me. He held me tight to keep me from face planting right into the grass. When we approached the clearing, he stopped then let me go.

I stood there on my own power but swayed a bit. I giggled at that. He watched me a moment appearing concerned. I took a step and stumbled. He caught me in his arms. I trembled all over with much sickness in my stomach. That was not expected. I frowned at that most unpleasant feeling.

"Casper, I thought dead people wouldn't have to hurt or be sick anymore. I do feel the foulness of my insides. Does this stop in time or am I doomed to be in agony like I was in life for all time?" I shook harder feeling more ill by the moment.

Casper came toward me and pulled me into a hug. He stroked my back appearing to be empathetic for my dismay. I felt him kiss the top of my head with that disfigured face of his. I began to weep thinking that he was telling me this horrible sensation would never pass.

He heard my sobbing and pulled back. Casper reached out and stroked my cheek with gentleness. Then he used his sleeves to wipe away my tears. I stood there shivering with the rising illness as he run his hand across my chest tracing several of my straight razor scars. I heard him let out a sigh just before pulling me back into another loving embrace.

I snuggled close to him trying to borrow his warmth. I had become aware that I was freezing. Death was beginning to scare me. It was as terrible as being alive ever thought to be. Casper felt me shaking from the cold.

He took off his coat and helped me put it on. Then he held me another few moments to be sure I was warming up.

I sniffed back my crying. "Thank you for being so good to me, Casper. I want you to know no matter where they take us you can depend on Mad Maxx to be your friend." I felt Casper rub my back in approval.

Then he pulled back and used the hand gesture for me to follow him in silence. I nodded that I understood. I was glad he knew all the Dominant sign language for commands. At least I would know what my new friend wanted to say from time to time.

I followed behind him as he walked toward the fading Haus. I staggered behind him doing my best to keep up. I stumbled several times but didn't fall. Casper would halt and watch to be sure I had regained my footing.

When he was sure I was okay, he would resume our trip. I gasped when he opened the exit door that was next to the back stairs for the Voting Council, Elders, and Head of the Haus. I was impressed this bank robber knew so much about that hell hole. I figured he was haunting the place because the living residents were evil enough in deeds it made him feel better about the ones he did when alive.

He pulled me inside the short hallway and shut the door behind us. I leaned onto the wall, panting from fatigue. That brisk walk had nearly wiped me out. My stomach was rolling, and the world was rocking. I felt like I was on a ship with sea sickness.

I hugged myself tightly trying not to vomit while Casper peeked around the corner to see if anyone was around. He came back and watched me turning green with nausea. He reached out and ran his fingers through my sweaty hair.

I looked up at him in mild discomfort. "I think I may throw up, Casper. Is this normal for the dead, to feel so sick?"

He nodded then reached out and grabbed a handful of my hair on the back of my head. I gasped for air when he tilted my face to the sky. I thought maybe this was a remedy for my upset gut. I didn't fight him until he leaned down, lifted the bottom of his kerchief, and forced his lips to my own.

Casper roughly held me to the wall with his free arm and my head still with his handful of my mane. I tried to push him off me, but he was too strong to budge. Weakness began to fill my limbs. I whimpered and held my mouth shut with all my might. He licked and pushed against my face with much vigor.

When I still didn't comply with his interest in the French kissing, Casper let go of my chest and grabbed my jaw. I groaned as he squeezed the sides of it with much strength. This forced me to open my mouth slightly. He ran his eager tongue inside. His hold prevented me from biting or closing it against him as I had been.

I became dizzy to the point of confusion with my head aching like it would blow off. Casper's kissing became more heated. He let go my hair and began pawing me all over. I felt my knees begin to buckle from my weight that seemed increasingly heavy.

I tried to speak, though it was impossible with Casper's tongue in my mouth. "Stop, please Casper. I don't want this." I managed to sputter out just as I felt the ghost grab my cock.

My eyes rolled back into my head, and I fell to limpness. Casper held me up by my head and crotch. I was unable to move. I was so fatigued I would nod off then awaken from Caspers rough handling unsure where I was or what was happening.

My sudden drifting away caused Casper to cut out his heavy petting. I felt him lift me up into his arms. He threw me over his shoulder like I was light as cotton. He walked up the stairs as I faded in and out of consciousness. I thought I was flying up those steps, but nothing was making much sense. I began to believe I was dreaming the whole thing.

I saw Casper as he gently pulled me from his shoulder and laid me against a wall sitting up. He reached out and removed his jacket from my naked chest with much care to not injure me. I blinked at him with slow thinking and lack of recall for the situation.

All I could remember at that moment was that Casper was my friend. I smiled at him while he put his coat back on. He looked around to see if anyone was looking. I tried to see what had him so anxious but then I forgot why I was peering down the hallway like that.

I turned my slow attention back to my friend. "Thank you, Casper. I see you in the Great Hall for our haunting this Haus after I get a little nap, ja? Can you tell my lambs I am sorry I missed their breakfast? I guess I failed them. They will be here in the land of the dead soon. The Dominants eat everything. Did you know they murdered me? Not nice of them but then dead is better than I thought it would be. Did you say your name is Casper? I want to be your friend, but you must stop kissing me. I don't like you like that." I nodded off for a second there, I think.

He nodded then took off with stealth and speed for the back staircase. I watched him already forgetting what the hell I was looking at. I closed my eyes hoping rest would clear my head of the fog. It seemed I should know where the hell I was sitting but at that moment, all that mattered was sleep.

I awoke to Leo's face floating above me. I thought I was dreaming again, so I closed my eyes to go back to sleep. He shook me with vigor. I groaned out telling him that it wasn't morning yet, but I would get his coffee in a moment. I was just so tired.

"Christian Axel, wake up, meine heart. Where have you been? Honey, you need to wake up. You are in the hallway. You are napping by the door. Why did you break out of the apartment?" I heard Leo saying but he sounded far away.

I opened my eyes to see him looking more concerned than ever. "Leave me be, Leo. I told you I will get your coffee in a moment. See the sun isn't even, wait, what the hell?" I sat up feeling stiff and groggy but suddenly aware I was not dreaming this shit.

I looked down the hallway and then back to Leo. "I uhm, was looking for you. Are the lambs, okay?" I was unsure of what exactly was going on, but my memory was flooded with images that made my blood run cold in my veins.

Leo shook his head. "I went out and let you sleep in meine Spatz. I come home to find the door wide open with my screwdriver laying in the floor. You have been missing for over three hours. The whole fucking Haus is looking for you."

I winced and rubbed my sore eyes noticing my flesh ached with the feeling of a brutal coupling. "I was searching the Haus for you Leo. When I woke up you were gone. I missed feeding my lambs and got upset. I went looking for you everywhere. No one seemed to know where you were either. I gave up searching hoping you would eventually come back. I guess fell asleep by the door waiting for you. Maybe I had another seizure or something?"

I had suddenly recalled the events of that morning after I broke out of the apartment. It turned out that fucking Julius was the rapist. Well, he found out the hard way that the Priceless silver collar is not the compliant slave he thought him to be.

Satisfaction filled me as I thought of him floating with the bones of those he murdered in the well. Then I shuddered when I remembered that killing him had only half solved my issue. It seemed there had been a second rapist, but was there really?

I let Leo help me to my feet wincing at the pain on my hindside. I had definitely been viciously violated, but by who? I started wondering if maybe the sensation of being dry raped was from Julius and I imagined the second man. I could remember having to give him a blow job. I realized there was the possibility he had sodomized me as well, but my mind didn't record it correctly.

You see, Meine Liebe, I often have trouble understanding events that happen in a linear way. I was aware that Julius could have completed his rape before I hit him. That I got that mixed up believing it happened once he was already dead. With that trauma over Christian's brutal murdering of him I imagined it likely I only thought there was a second rapist.

It was too crazy, even for me, to believe that there were two of these cocksuckers using the same techniques to have their way with me. The evidence against it was that I was still alive. If that second guy was not just a hallucination there was no way he would let me live. I had killed his partner. He was out there in the forest watching when it happened.

It simply made no sense. No one in their right mind would watch me dump Julius in the well without trying to stop me. I was supposed to believe that this guy not only ignored my murder, but he stalked me, raped my ass, made sure I didn't freeze to death, walked me past the guards then dumped me by Leo's door still breathing. Yeah, not likely.

I sighed with relief that I had gotten the sorry sonofabitch that had been hurting me at last. I was finally safe from Julius's sexual assaults. There was no reason for me to be stressing over imaginary partner rapists sneaking around in those woods.

It was a drug-driven nightmare and that was the end of it. It was clear to me that orange drink Leo gave me had some ingredient which caused my dreams of being a Spatz (sparrow) in Leo's bed, falling asleep in hallways and weird visions of rapists named Casper.

I stood up with a bit of dizziness and followed the anxious Leo back inside our home. He quickly shut the door and bolted all his locks. I stumbled to the couch and dropped to a knee feeling sick, groggy and like I desperately needed a shower.

Leo stood there staring at me with an expression of fear. "Meine Spatz what the hell am I going to do with you? I cannot leave this apartment for a second without you panicking, then breaking out. You are never to leave our haus again without my permission, Christian Axel. It is not safe for the Priceless to be wandering around this place alone. You are not well, meine heart. This is a fourth directive. Now I am the liar because I told you I make no more than the three directives. What a fucking nightmare this has been trying to keep you from injuring yourself." He sat down in his armchair rubbing his forehead, appearing most upset.

I kept my eyes to the carpet feeling bad that I made him feel angry at me. "I apologize, Leo. I won't go out anymore without your say so. I don't want to go out there anymore anyway. Can you call and get Rudolf to feed my lambs?" I shivered feeling afraid of the idea of leaving that apartment ever again after the morning I had.

Leo startled then glared at me. "What? You wish to shirk your duties to the lambs? Christian Axel, those creatures are your responsibility, not Rudolf's."

I winced and nodded. "Ja, you are right. I apologize again Leo. I beg you to tell me what to say to make you happy with me like you used to be. Maybe it is time to thud me? Will that make you feel more comfortable? I am sure I deserve it."

Leo groaned then rubbed his head roughly. "Gott damn it. I don't want to beat you, Christian Axel. I want you to be well. You are kneeling again, even though I keep telling you not to. You run off when I go to town to get more books. You scour your flesh when I turn my back. I am at my wits end to know what to do to help you."

I shrugged. "I don't have anything of worth to offer you in return for my short comings, Leo. I can offer the special services to make up for the trouble and money I cost you. I will accept the thud for my misbehaving. What else is there I can do for you? I have no other abilities nor things of worth. You already saw fit to confiscate my razor and sandpaper for your own. All my other treasures are where I cannot reach them to pay you for what trouble I have caused you."

Leo gasped. "Meine Spatz, I have money and things galore. I don't need another fucking material thing in the world. What I want from you is your love. That is the one thing that no money can buy. It is supposed to be free."

I looked up at him with an expression of hatred. "I beg to differ, Leo. Love is the most expensive thing in the Gott damned universe. Only the wealthiest of hearts can afford to pay the down payment and the payments are hefty. Many end up getting it repossessed thanks to their inability to maintain that money pit emotion. The fools rush in Leo, but the wise are cautious. If you thought loving me was going to be a snap because I was the boy without options and would be an easy mark, better think again. Even the beggar can choose to starve rather than take the garbage another passes him with a mislabel that he is getting the finest feast."

Leo gasped. "Are you saying you don't love me for truth, Christian Axel? I thought you did. I love you so much I go out of my way to try to help you by granting you freedom, and pets, and…" I interrupted him.

"You want to love me, Leo? Then do so without all the bullshit lies. You, my love, are my Master. I am your submissive. I would run away from you if I could. It is nothing personal, but I didn't choose this life, Leo. You keep me here saying for me run would result in my death. Well, I noticed you didn't ask me what I wanted. Maybe I would prefer to die free then be trapped in this fucking Haus wearing this Gott damned collar. You can grant me mercy of an orgasm during the special services, but I don't want to be fucked by anyone, ever. I want to be free. I want to have the right to choose where I get to live and who I fuck, like you do. You say don't kneel, do what you want in our home, but you take my things, lock that fucking door, slip things into my orange juice, and even tell me how many showers I can take a day. I can do whatever I want like the freeman but only if you approve. You try to half-ass undo the training that I suffered many a thud to learn, and then hand me over to other men that expect that I kneel and let them fuck me. That is not love, Leo. That is, you are making yourself feel better about the subjection that you are just as responsible for."

Leo sighed. "I thought we agreed that I would take the punishment and pay the price for my part in Peter's plot. Christian Axel I have been doing both. Surely you cannot deny that."

I laughed with much bitterness. "I am not talking about that shit with Peter neither. I am speaking of right this fucking moment. What if I say to you, I want you to take me from this Haus, cut off this collar and drop me off in the world right this minute. What do you say to that?"

He looked at the floor. "I cannot do that meine heart. You would die from starvation if the Guard didn't get to you first."

I nodded. "Ja, just what I expected you to say. I tell you what I want, and you say nein then make excuses for ignoring what I desire."

Leo teared up. "I love you, Christian Axel. I cannot just cut you lose to be killed or die. That request is totally unfair to use when assessing if my love for you is real."

That made me chuckle. "There is no doubt you love me, Leo. You love what I can do for you, how I make you feel. That is okay. I was trained to do that for you. It is time to clear this misunderstanding between us. You must stop acting like I have a choice, and I will stop distrusting you. You tell me what you fucking want from me, and I will no longer be afraid you are going to tire of the games you play with me and yourself. I am not Maus, and you are not my brother. I am also not a love seeking character in your romance novels. I am a fucking hostage. If you toss my collar, I will be passed on, then abused to death. You will live on and love again. I am happy to accept the love of my Dominant Leo as it is. I can even love my Master back width the understanding it has conditions."

He sniffed back his tears. "Our love has conditions. What are they? Tell me and I will meet them. I want this relationship to work, meine Spatz. You tell me how to help you and I will do anything to make you as comfortable as possible. There is nothing in the world I want more than to hold your heart for my own the way you have taken mine."

I smiled with much sadness filling me at the rough truth of this matter. "The answer is simple, Leo. You need only be my loving Master. I will then be your grateful, loyal, and devoted submissive. Otherwise, you love me without condition by taking me the fuck out of this hell hole and let me go. There is no middle ground in this brutal world you choose to be a part of, and I cannot escape. You must make this decision, because Leo, I am denied such a right. Do you understand now?"

He nodded while looking at his feet. "Ja, I see what you are saying to me, meine Spatz. Does this sad truth mean you don't see my home as your own, nor want the lambs or hound?"

I got up and dropped to his feet in a kneel. "I see this as the only home I have ever been offered, Leo. I thank you for the lambs and Der Makellos. I want you to know I do love you more than you know for being a good heart to care for me so much. I give you something that I never gave to anyone before you. You have my ardent desire to serve your needs with peace in my soul. Master Peter, Jonas, Claus, and Bladrick force my services. Not my Leo. I beg him to allow me the thrill of bringing him joy and thank him for the pleasure of it. For Mad Maxx the Brutal

Priceless collar this is as close to love as he can ever get for one that subjugates him and takes his right to choose." I smiled at him with as much kindness as possible given the situation.

Leo reached out stroking my cheek softly with tears beginning to stream down his face. "I accept the kind of love you offer me with gratitude, meine heart. I apologize for being the blind fool. You are right, you are not Maus, nor my free lover who chose me. I didn't want to admit that I am the criminal for abusing the sweet boy that doesn't deserve what he has to endure. Thank you for your patience and correcting my ignorant behaviors."

I grabbed his hand from my face and kissed it. "I thank you for your fairness in treatment and mercy, Master. I will never fail you in my perfect service to your pleasures. I give you my word. I further promise that when this metal is broken, I will never abandon my Leo no matter where I go or what I do. I swear you will always be welcome in my heart and home just as you have done for me in my darkest hours. I can never forget you cared for this worthless boy when no one else ever bothered."

Leo pulled me to his lap and showered me with adoring kissing in an affectionate embrace. Our strong bond was finally completed. He understood that the feelings we had for each other were not going to result in a typical vanilla love affair.

This was to become a true D/s relationship forged from equal respect, mutual understanding of our roles, and deepest trust in each other. Leo would become the rudder on my ship that was adrift in the tempest driven sea. For his loyal, loving, fair service, I would be willing to give up my own life to defend him and overlook my own nature and comfort to fulfill his darkest desires. Leo had become the benign Master that every submissive dream of, but few ever get lucky enough to obtain.

When Leo's adoration became heated with much lust, I pulled from his hugging with a frown. "Master I must get a shower before attending to your interests. I feel foul. There is also preparation that must be done to meet your needs properly."

He nodded. "Ja, I understand, meine Spatz. I release you to attend to what you must. I see you soon?"

I smiled. "I assume this would be your desire, if so, the answer is always ja. I must ask you what of the lambs. Master? They must be fed. I overslept."

He kissed my cheek. "I called Rudolf to attend them this morning. We can feed them their dinner later. I wish to spend some time just the two of us this afternoon. Jonas will be back soon to take you from me. I intend to use all my time left to the best of my ability."

I groaned. "Ja, Master Jonas will be back the day after tomorrow. I think that will be rough with his lust unattended two weeks like it has been. I may need some of that mystical orange juice of yours to get through his cycle days." I chuckled at that.

Leo laughed hard. "You are a gem, meine Spatz. I don't know that I could joke about that cruel Jonas if I were in your boots. You never cease to amaze me. Your inner strength is a marvel."

I jumped off his lap headed to that shower but looked back with a smile. "I am Mad Maxx the Brutal, Master. What else could one expect but the unbreakable spirit, ja?" Leo howled with laughter behind me as I rushed to the spare bathroom.

I turned on the hot water waiting for it to steam up the mirror. I grabbed the blue bottle of magical liquid. I then braced myself for the agony of the scalding heat. I stepped inside and felt immediate relief as the stinging droplets seared away the foulness of Julius's violations of my flesh.

I scrubbed with fury at the healing skin on my legs and backside. The crud of disease was festering in every cell my waist and below. When the soap did nothing to break free the claws of my sickness, I snatched up the blue bottle. I began to spray feeling near mad with fear. I could feel that nasty stuff's knifelike teeth hanging on to the boy with all its might.

I used half the bottle to end this invasion. I realized it took twice as long to rid myself of the illness as the day before. With a panic I discovered the disease was mutating. It was figuring out how to beat the magical blue water cure. At that rate, I only had a few days before it would take the whole pool of this stuff to combat my growing sickness.

I was freaking out pretty bad till I thought of Master Jonas. He had sandpaper in his apartment. He would never stand for my not remaining clean shaven everywhere but my head. He was just like Peter, demanding not a single strand of hair remain anywhere on the boy otherwise.

The Vampire would be home in two days. I was sure I could hold off the disease with the blue water cure at least that long, if Leo got me more of it. I calmed myself with the promise I would ask Master Jonas for another razor. Then it would be easy enough to snatch a fresh piece of that wonderful sandpaper.

I sighed in relief while finishing killing the last of that nasty crap. I had a good plan for keeping it from consuming the flesh. All would be alright for a change. I heard the bathroom door open. I put down the bottle wondering if Master Leo had come to check on my scouring.

He pulled back the curtain. I saw him smiling with wickedness. I let out my breath realizing that he had come to take his rights for the special services. I backed up so he could enter the tub without hesitation. This kind of task for my Master was my job after all, ja?

Leo stepped in and let out a blood curdling wail. "Gott damn, that is fucking hot. Christian Axel, turn on the cold water." He jumped from the tub as if shot while rubbing the streams of water off his skin still yelping loudly.

I winced but leaned forward and worked the facet until the shower was a temperature the way he seemed to enjoy it. "I apologize Master. I didn't expect you would wish to join me. Forgive my stupidity for not warning you before you tried to get in."

He glared at me with suspiciousness and stuck his hand into the stream. "I thought I told you the blue water cure works without this boiling yourself half to death. You will not run the water in your showers to that horrid temperature anymore. That is a directive, Christian Axel. I may wish at any time to join you in the bath. You will make sure to always make it comfortable for me." He got into the tub appearing satisfied that I had made it hospitable for him.

Leo didn't waste time asking for his rights. He knew what he wanted and was not coy about it as in the past. I attended his demands for the special services without quarrel. As in all the other encounters with this Master, I found interest in his gentle, loving touch. He rewarded me with my own release for my sacrifice of allowing his penetrations with orgasm within the boy.

When both of us were spent of our lust, I cleaned him with grace and skill. Leo got out and I dried and I dressed him in a bathrobe. He left me to attend my own bathing of my latest soiling. I used the rest of the blue water to clear the disgusting sickness that had already begun to take ahold of my flesh. I wept while drying off feeling quite helpless over this never-ending battle to stay fresh. I believed sooner or later it would eat me alive.

I dressed in a blouse and fresh breeches. I didn't bother with my boots this time. I took the empty spray bottle then went to the living area to seek out my Master. He frowned when I dropped to a kneel and put the thing on his coffee table without saying a word.

"You went through the whole bottle in less than twenty-four hours, meine Spatz. Is this illness of yours that fast or is there merely a lot of area to cover?" He narrowed his eyes looking at the bottle.

I shook my head. "Both, Master. Forgive me for costing you a fortune but I must have more of it before the morning. I will need another shower soon enough I think." I dropped my gaze feeling sheepish to be so damned expensive.

He sighed. "That is okay meine Spatz. I will call the mystics and get them to send a new batch plus a backup bottle as well. When Jonas gets here it will be difficult for me to know when you have run out. I will make sure to keep an extra bottle here if you need it in a hurry. All you need to do is knock and say I need the cure. I will hand it to you so you can grab it and run, ja?" He reached out and grabbed my upper arm pulling me to sit with him on his sofa.

I nodded. "I am most grateful for the mercy of it Master. When do we go to feed the lambs?"

Leo pulled my head to his chest and rubbed my arms. "Not for a few more hours. I want to just lay here and snuggle while. You haven't even left yet and already I feel my heart growing heavy with sadness over it. I never realized how lonely I really was until you came into my life, meine Spatz. I want you to know how happy and complete you make me feel." He stroked my hair looking at me with much adoration.

I smiled bitterly at him. "That is what I am supposed to do, Master. I hope you don't think my love for you is less because it is not the romantic kind you hoped for. I know you do not have to love me to make me fulfill your pleasures, but you do. For that I am beyond grateful. You are all I can think of day and night. I grieve when I leave your bed and our home. I want nothing but to see you thrilled. My love for you is a stronger love then the ones your fantasy books discuss."

He laughed out loud. "You know I am wondering if there is truth to the saying madness borders genius. Meine Spatz, sometimes you say the wisest things I ever heard in my whole useless life. I know another lover can say they would kill to keep me safe, but they would be only saying such a thing thinking it romantic. You on the other hand, if you say you will defend me to the death, I know it is the truth. I never have to pretend I am something I am not, nor apologize for being less than perfect. You accept me as I am and in fact demand I be myself without all the bullshit acting. That alone is worth everything in the world to me."

I chuckled at that. "Good thing, Master, because if that blue water is expensive, you better get out your check book or take out a huge loan. I will need an ocean of the stuff."

He pushed in my nose and made a silly face at me. "You don't worry about the money I spend. I do believe that is my role as the Master. If I couldn't afford my

Priceless metal, then I shouldn't have accepted the responsibility, ja? I will make a call to the mystics before we leave for the lambs. It will be here with our dinner when we get back. I want you to drink some more of the preventative before bedtime tonight too. No arguments. That is my pleasure, and I do believe you wear my collar." He smiled with happiness as I rolled my eyes at that.

"As you wish Master. Can I ask what the hell they put in that orange crap? I have had the strangest dreams from it." I wrinkled my nose and smacked my lips with disgust at the memory of its foul taste.

He pulled me into a tight hug. "If I knew meine Spatz, then it wouldn't be magic now, would it? You drink it and keep your complaints to yourself. I want to read my novel to you and enjoy your company. Be still and listen. This discussion is over."

I nodded that I understood. His acceptance of his role was alleviating my confusion while on his leash. I didn't always agree or even like his commands, but at least I knew what he wanted from me. I was glad to not have to always be guessing or trying to decipher his underlying desires as I had been.

Thanks to this, I believed I could trust what he told me as truth. I knew Leo really did wish the best for me. Well, as long as it didn't leave him without my company for good. He would do anything possible that I asked, but the one thing I wanted more than anything else, my freedom from the submissive's silver burden.

It was simply too good a deal for him or any of my Masters to let me go. As long as I wore that fucking collar, he had full authority to hold me there to obey his every wish and unable to leave him. I understood gaining my will back was something I would have to do myself. No one, not even my beloved Master Leo was going to do it for me.

He had me put my head on his lap like he was a pillow and lay across the couch. I laid there quietly while he read his romance novel aloud to me. It didn't take long for me to nod off in a peaceful, dreamless sleep. I felt safe with Master Leo. after the happenings earlier, the mercy of lower anxiety is something I needed more than anything.

Leo woke me a few hours before dark. He had me dress in my boots and coat, then he took my leash and led me from the apartment. We walked along in silence headed for the barn to feed my lambs. I saw Olaf and Vilber standing at the door as usual. I shot them a look of pure hatred.

Their eyes went wide in pure horror when they saw me walking in my protocol behind Master Leo. He stopped at the door to speak pleasantly to the brutes. He had no idea these two had near killed his collar earlier that day.

I smiled with evil as they shot me looks of disbelief. Neither could have imagined that I not only survived their assassination attempt but had made it back into the Haus without alerting anyone to their cruel behaviors.

I lifted my finger and used my thumb to motion the pulling of a trigger on a gun. I closed one eye pointed my "pistol" at each of them and pretended to shoot while my Master prattled on about the weather to dem.

The brutes trembled with terror. They assumed I was willing to tell on them, but they were mistaken. I had no intension of the Voting Council getting the pleasure of sending them to their graves. Nope, that thrill would be mine exclusively. Vilber and Olaf were on my shit list, and I wouldn't stop till their names were marked off with a big X.

I followed Master Leo out and heard Vilber say under his breath, "A fucking demon. I knew it. Gott damned spawned right from the loins of hell. You cannot kill the damned, Olaf."

I grinned that he thought that of me. I was happy to call the devil my father. Sure, beat the fuck out of the real one I had here on the Earth. That brought Peter to my mind. As I traveled to attend my little ones, I wondered why he had been hell bent to chase me. He seemed beyond angered.

His intense irritation made no sense. I had not even seen him in days. I couldn't have insulted him and the other things that pissed him off, the spitting and ignoring him in the Great Hall, I thought he had let slide.

If he had been looking to punish me for that shit, then why did he let me go in the dungeon the day I seduced Mistress Heidi. I shrugged to myself deciding that my old man likely wanted to just be a dick like he always is to me.

He saw me and thought it was an opportunity to lord his Dominance is all, right?

I found my lambs eager to get their loving and grain. I stood there watching them eat with thrill. Master Leo stayed at the side of the pen keeping his eyes on me. I was grateful he didn't wander. I was feeling a bit nervous being outside again. I no longer cared for the vastness of this open space of the Haus yard. Too many places to hide. It made it easy for evil people to sneak up on a Priceless collar, you know.

I stood there ready to grant my lamb children their instructions for my absence when Master Jonas came back. I waited for them to finish their meal so I could be sure they all heard my directives. When at last Ryker stopped lipping around their feed trough and came to join his brothers and sisters, I gave them the sad news.

"Now you all listen to me. I have many important things to attend in the near future. You must all mind your manners and remember your places. There are wolves everywhere waiting to eat all you for their dinner. I must insist you do as I tell you or there will be consequences you will not appreciate. Ryker, you will stop goading Milo to try escaping this pen. Geraldine, you mustn't keep pushing Annette down. If she busts her knees, she will be of no use to anyone. They all expect quick kneeling and perfect service you know. You all need to stop behaving with such insolence toward your betters. The sound that you hear of electricity, which is the carving knives of the Dominants. Do you listen to me or only pretend to care? You cannot grow up with such attitudes of destruction in your brain. Gratitude is a paramount key to all life. I insist you take that to heart and follow my directives. Stay away from the woods. Casper lives there and he robs banks. He has no face, but I bet he likes to eat veal, ja? If he catches you then you will be eaten alive, then thrown into a hole. I will see all you in the morning. I don't mean to scare you, but this world is a brutal one. There is no escape for you. Good night little ones." I turned around and went to the gate waiting for Master Leo to open it.

He stood there with a stunned look on his face. I was unsure why he stalled at letting me out. I began to tremble thinking he intended to cage me with the lambs. I knelt, frightened out of my mind.

Master Leo cleared his throat. "Christian Axel, get off the ground. This that you told the lambs, uhm, never mind. I let you out. You follow me back to the Haus and remember no trying to run or break out of my door locks." He opened the cage, and I rushed out grateful he had changed his mind about abandoning me in the barn.

He took off and I followed in silence wiping the sweat of my panic from my forehead. I noticed he was headed to the playground. I gasped but said nothing. I felt my heart speeding up as I watched the dark threatening forest seem to grow wider and taller.

I thought for sure I was going to have a panic attack if even a bird called out, I was so highly stressed. I didn't want to be anywhere near that place for the rest of my life.

Master Leo stopped at the swing set. "Here meine Spatz, I will swing you for a while, ja?"

I shook my head, never taking my eyes off those foreboding trees. "I take the punishment, Master. Thank you for the mercy."

He startled then looked at me with curiosity. "You take punishment rather than doing something fun? What the hell, Christian Axel? I thought you enjoyed the swing set."

I backed away from the equipment, openly shaking in fear. "Nein, I would request you take me back to our apartment. You can punish me there. I don't wish to swing, not today, not ever. Thank you for the mercy of it." I could see the shadows in the trees moving and hear the howling of the lost souls that lived in that well screeching in my ears.

Master Leo stood there in disbelief. "I am not taking you anywhere till you tell me why you suddenly don't wish to swing anymore."

The wailing of the dead silvers was becoming deafening. "Please, I beg of you Master. I don't want to be here anymore. Take me away from this place." I fell to my knees covering my ears and trying to roll into a ball filled with paralyzing terror.

Master Leo came at me and grabbed my upper arms trying to lift me to my feet. "Get off the fucking ground, Christian Axel. What the hell is going on with you? Tell me," he jelled.

I screamed with horror as I saw what stood at mouth of the woods. Julius stood there holding the leashes of the children from the well. They were all bloated, blue, and with their heads tilted to unnatural angles. Julius looked like sausage meat from the shoulders up.

His laughter filled my ears with an earth-shattering sound. The world tilted and bucked. I shrieked repeatedly as I broke Master Leo's hold to run from my undead attackers.

Master Leo grabbed my leash holding tight digging in his feet to keep me from escaping. I hit the end of it and nearly fell from the force. I wailed and pulled with all my strength covering my ears trying to drown out Julius's threats to rape me and throw me into the well.

I saw Peter and Malfred come out of nowhere. They ran over to aid Master Leo in holding my chain. I really freaked out at that point. I struggled and jerked at it hard. The three big men held me hopelessly at the end of my restraint.

Peter then let go of their end and came flying at me. I punched and kicked him as he picked me up and threw me over his shoulder. He didn't appear to even notice my blows. I kicked and flailed yelling with every breath begging for help.

Peter spun around to Malfred and Master Leo. "One of you restrain his fucking arms and legs, Gott damn it. This fit must stop now before someone sees it. Malfred, if you have rope on you, use it. Leo, you find something to gag his screams. Hurry the fuck up."

It took all the big men, but they finally managed to bond my wrists and ankles. Master Leo used his handkerchief to silence my wild screaming. I wriggled on the ground helpless to escape them while the men rested from their battle.

Master Leo looked at Peter, who was panting with his arms leaned on his knees. "Thank you for the help, Peter. I don't know what the hell happened. He just freaked out."

Peter growled. "I tell you the boy is much stronger than I recall. What the fuck are you feeding him? I didn't come to help you, Leo. I am protecting my investment. That boy will belong to me soon enough. If I hadn't stepped in your weak Dominance will get him shot by the fucking Guard. What the hell are you doing bringing him outside in the first fucking place? He is too valuable to risk like this. You seem to not understand if you turn your back a second, he will kill you, then run off. Your love for this boy has made you blind."

Master Leo scoffed. "Fuck you, Peter. You will not tell me how to attend my own Gott damned submissive. I will take him anywhere I want. You speak to me like you are my better. Well guess what? I am above you now, cocksucker."

Peter nodded with a bitter smile. "Ah, that is true, for now. You better watch who you make into your enemies,

Leo. That boy belongs under lock and key, and you know it. One day Malfred and I won't be here to help you."

Master Leo snorted "Ja, I would ask you to keep your advice to yourself. Thank you Malfred for the help as well. Do you want to insult me like Peter?"

Malfred shook his head. "Nein Leo. I accept your thanks with grace. However, Peter may have a point. This boy is far too precious to risk removing him from the safety of the Haus. If the Priceless were mine, I would never let him out of my sight or out of my apartment. There are too many out there looking to have a taste of this boy. Take it from one who knows. You hold something this desirable, eventually another will steal it. My Tamina is long gone and not a day passes I don't feel robbed of the joy she brought to me. There was never a more desired silver, until this treasure I saw before me. You are the luckiest in the Haus with her gone. If you want to stay that way, better keep this boy caged and locked away from sight of those that will be tempted to possess it for their own."

Master Leo shot a look of hatred at Peter. "See that is the gentleman's way to caution another Peter. Malfred has manners. I will take your wise Council into consideration Malfred, and I thank you for your kind words. I need to get this collar inside before it gets dark. I don't want the fucking Haus seeing this drama though." He looked at the Haus with frustration in his expression.

Malfred chuckled. "Take him in the exit door behind the Elder's stairwell silly. No one messes around there.

You will have privacy to get him home without tongues wagging."

Peter nodded. "Ja, let me help you, Leo. I wouldn't want this wiry bastard to escape you yet again." He came at me and threw me back over his shoulder.

I wiggled and groaned loudly but was unable to do shit about his rough handling. Master Leo and Malfred followed behind us doing their best to block the view from any that may pass this strange group.

Malfred and Master Leo spoke in low tones about Malfred's loss of his expensive metal Tamina. I recalled that name belonged to the silver Olaf had requested for his favor to swear allegiance to me during Peter's reign. I was sorry to hear that the brute managed to get that poor girl pregnant. Malfred had loved her so much, he sold her silver to another Haus rather than put her to the circuit.

You see in the Haus getting pregnant by your Dominant is no crime. It is often a happy occasion and celebrated by both the submissive and their Master. In most cases, there are always exceptions. That said, to carry the child of a black collar when you are silver is proof of the deadly crime of theft of services, even if approved by Master Claus.

Often the silver will refuse to name the father. In the case of Tamina, Malfred was proven infertile in his youth. When she became pregnant, a stool pigeon said that the father was a black collar but didn't name the culprit. Likely that brute Olaf bragged to some of his buddies of his taste of the most coveted silver collar female in all the Haus. That shit came back to haunt poor Tamina.

Malfred was called to the Council to answer. He had no choice but to admit he had allowed a leash with an unknown black collar. He was never told of the man's identity. They handed down the judgement that she be sent away since the Haus will not put an innocent unborn child to death.

Malfred did as was ordered with a heavy heart and much grief. It had been almost a year. Yet, he still spoke to Master Leo with tears in his eyes at his loss of the aging Tamina. I listened and felt bad for both him and his disgraced submissive.

He had not blood bonded her which would have brought him some leverage. He told my Master if he ever got lucky enough to possess another silver of such high quality, he would never let them go and bond them in his crimson the second he had possession.

I admit I felt guilty about this horrible situation. It was my fault for asking Master Claus to grant me the favor for Olaf. I had not considered how my plans to obtain the brute's loyalty may affect another that was innocent of my struggles. Like Ben, I never thought a second how my actions affected others nor thought of the helpless Tamina that was merely following her better's commands.

We made it into the Haus and up the stairs without being spotted by any of the residents. I stopped my struggling feeling mighty low over the fate of Tamina.

Then when we arrived at the apartment, I heard a familiar voice yell out.

"What the fuck. Peter, you put my man down, you motherfucker." Master Jonas was waiting at Master Leo's door. He was home early, Gott damn it.

Peter lowered me off his shoulder into the eager claws of the Vampire. He was pissed. I saw his eyes searching the group of them till they landed on Master Leo.

"Leo, why the hell is this cocksucker here? And Malfred? What the fuck. I thought I told you never to set foot on my Gott damned floor again. You both have gone too far this time. I have had it. You think you can just ignore me, molest my man, and get away unpunished for it. I think not. Leo are you in on this bullshit." Master Jonas held me tightly in a hug to him while he threatened the men.

Master Leo came forward. "Jonas calm down. Peter and Malfred were aiding me in subduing the boy. He went into a fit out by the playground." Master Jonas interrupted him.

"A fit out at the playground. Are you out of your fucking mind? What was Mad Maxx doing out there with you." He squeezed me even tighter causing me to cough and struggle for breath.

Malfred saw my difficulty. "Jonas, you are crushing the Priceless. Be careful with him. Calm the fuck down. Peter and I were just leaving. Come on brother, let's leave these grouchy Elders to attend their own quarrels. I for one will not stand here and be insulted when all we did was assist them to save what neither is worthy to possess." Peter nodded and the two of them tore off down the stairs leaving me and Master Leo to deal with the furious Vampire Jonas.

Master Jonas pulled the kerchief from my mouth demanding I keep silent. I coughed and gagged but didn't say a word nor scream out. Master Leo approached me with much worry in his expression.

"Mad Maxx what happened back there, my love? You were very upset by something. Did you see a hallucination?" He came closer and Master Jonas put up his arm preventing him from touching me.

"You back the fuck up, Leo. I'll tell you what happened. You took this ill boy outside. He is not used to such stress as the fucking playground. If you must take him out, you stroll only. Too much excitement will set off his symptoms," Master Jonas bellowed out.

Master Leo threw his hands on his hips then snapped in the air. "Oh? How the fuck does a swing set compare in stress level compared to your cutting him up like a Christmas ham, Jonas, prior to your drinking his blood? I don't see him throwing fits over that insanity."

Master Jonas glared at Master Leo with much hate. "You will mind your business about what happens with my man in our bed. As it is I am taking Christian Axel home now. I will find out what the hell happened later after the boy has time to calm down. I would be freaking out too if that nasty Peter and vengeful Malfred had me tied up like a hog. Get the fuck out of my way." Master Jonas lifted me like the bride storming back toward his apartment.

I looked behind me at my the obviously grieving Master Leo. Master Jonas didn't even let me get my things nor kiss Master Leo goodbye. I was helpless to struggle with my wrists and ankles still bound. I whimpered as the still furious Vampire hauled me inside his apartment.

He shut the door never dropping me from his hold. I began to tremble as he headed down his hallway right for the bedroom. I closed my eyes vishing that my heart would stop before he got me to his bed and rough lustful coupling.

Of course, I didn't die. Master Jonas reached his bed and tossed me onto it. He put up his keys then came back my way and crawled up next to me. I looked away as he stared at me with a smile.

"Did you miss me, my love? I couldn't stop thinking of you. I know you are upset with me for leaving you with that bastard Leo, but I am here now. Those men didn't hurt you, did they? Peter and Malfred minded their manners?" He reached out and stroked my cheek.

I shook with fear and continued to avoid his gaze. "I am not upset with you Master. You did what you thought best. Peter and Malfred didn't molest me. They only bonded me to calm my insolence. I had that coming. Are you going to punish me for acting a fool with Master Leo?" I was careful not to admit that Master Leo and I were not the enemies he thought us to be.

He shook his head smiling with all his pointy teeth. "Of course, I will not punish you for standing up for yourself against those bullies. That Malfred is the biggest toad with all his demands for my man's leash. He thinks he is owed a regular night with the Elders' Priceless because of his stupidly leashing that old silver of his to some black collar scumbag. Then Peter, well we need not speak of that trash in our sacred bed, ja? Did Leo treat you decently at least? Or was he cold as usual?"

I sighed. "I endured him, Master. Thank you for your concern but I cannot discuss this any further. Like it or not, he is my Master too."

Master Jonas grabbed the side of my head by my hair and pulled me into his forceful kissing. I closed my eyes and bit back my urge to scream that he ceases this unwanted and unwelcome attention. This is my job, and I knew better than to deny a Master. But somehow, it was harder than it ever had been before and that is saying something.

He began to pant then dragged me closer to him running his hands across my bonded arms and chest. I held my breath praying he would decide to wait to call in his special services. I wasn't ready. I had no blue water, and well, I didn't want him fucking me to be brutally honest.

To my despair I could feel his cock pressing into my stomach as he licked and sucked on my face and neck. I braced myself for the promise that he would not let that hard on go unsated. He reached down and grabbed my own manhood, shit. I forgot I wasn't wearing my chastity device. Oops!

The Vampire let out a growl that sounded like the angry mother bear. "Why are you not tethered. Who removed my chastity from you. Speak now, Gott damn it, or I will beat you half to death." He backhanded me so fiercely across the left side of my face I nearly went unconscious from his wrath.

I sputtered and fumbled for the words, feeling my face burning like fire where he struck. "Mercy Master. I removed it. I stole a hex wrench when Master Claus was napping. It was chaffing my hodensack. Please, I beg of you to forgive me." I braced for his next blow which was to my left ear.

"I don't care if that fucking thing is rotting your cock off, you don't remove it without my say so. You hear me?" He grabbed my leash near my collar and held me to his face while I gasped in terror unable to even defend my face from his attack or push him away.

I nodded. "It won't happen again, I swear it. Mercy please, Master." I began to sob and plead with all I had, praying he didn't hit me anymore.

Master Jonas dropped me to the bed and straddled me. I whimpered still weeping in complete fright at his flashing dark eyes and sharp gnashing teeth. He ripped open my blouse. I wailed out for mercy as I watched him pull out a pocketknife from his waistcoat. I thought he was going to cut my throat at last.

Instead, he used it to slice open my flesh of my left chest muscle. I screamed from the pain. Master Jonas smiled at my tears of agony. Then dropped his head and began to suck the blood welling up from the wound. I gasped and cried out, feeling the despair wash over me. I just wanted to go home. This was a horror I had neither expected nor been prepared to endure.

When he finished his blood drinking, he rolled me to my face. I did my best to quell my tears when he began to rip at my breeches. I realized with anguish he planned to dry sodomize me in his anger. I began to sob loudly when he violated me with his fingers to prepare me for his harsh entry. Just before he could start his intercourse a knocking began at the door.

I held my breath and shuddered with my tears still streaming as the Vampire held still listening. The knocking came again, louder this time. He cursed and told me he would be back to finish my punishment for removing the chastity device. The Vampire jumped off the bed leaving me there still weeping like a baby.

I pulled up into a ball waiting for his painful intercourse, praying he would cum quick to end my torments. I did my best to settle down but not even my masochist, Mad Maxx, could tolerate the dry penetration. All of us Max boys held still and braced for the coming horror.

To my surprise the Vampire returned within only a few moments telling me to come toward him. I did as commanded with much dread in my heart. I assumed he wanted a blow job to prepare him for his rights. I gathered as much spit as possible to try to get a bit of comfort while struggling in my bonds to reach him.

He lifted me to my feet and began untying my wrists. I stood there with a confused, but grateful expression. Master Jonas noticed my look.

He scoffed. "We will have to finish this punishment later. Claus and Bladrick have asked all the Elders to attend dinner with them in the Great Hall. They are celebrating my return. To deny them is bad manners. We leave in thirty minutes. You go to your room, change into something nice, then meet me in the living room. You have thirty minutes, Christian Axel. I have taken my blood feeding from you. Your intense fear of me will have to work this time, but I will need to figure out how to get the proper anger and lust one before my next in a fortnight. When we get home you have two choices. You come to me willingly and give me adoring service or I can finish what we started. Easy or rough, up to you, my love. Now get going. You are wasting time." He pointed to the door. He didn't have to say it twice. I fucking ran like the devil was chasing me.

I chose his favorite vampire outfit. I practically tore off my other outfit to change. That blouse I was wearing was ripped in half anyway. I rushed my makeup but noticed a huge purple and red handprint forming on my jaw. It took a lot of white cream and power to hide it. I winced when nothing I did hid my purple and pink left ear. That was gonna cause tongues to wag. I hated it when others saw I was punished for insolence. Not good for my reputation as the bad ass, ja?

I was feeling anxiety fill my every pore. I knew that Master Leo would be at the dinner too. I needed to see him to remind me that I had at least one sanctuary to help me endure this nightmare. He was my rock of stability and hope for a better day. Plus, he had my magic blue water.

I finished my assigned task with ten minutes to spare. I ran from the room with the gray door and rushed down the hallway. I passed the antique chair that Master Jonas had been working on and noticed he brought more sandpaper. It lay scattered about the floor in his library area.

I stopped my mad dash and slipped into the space. I didn't bother to mess around. I grabbed a sheet of that awesome stuff stuffing it into my inner coat pocket. I felt immediately relieved. I would not have to worry if Master Leo forgot to bring me the cure water. I had another way to beat the sickness safely in my possession.

I started to leave, and I noticed a strange looking knife with ragged teeth in his toolbox. I thought this would work maybe even better than the paper for scraping away diseased flesh. I grabbed it too and slipped it into my hiding spot. With a calmer heart I took off to kneel at Master Jonas's feet awaiting his lead to the Great Hall.

He smiled at me. "Ah, there is my love. You look better already. Now come, we will meet Claus at his apartment and walk down together. You need not worry as your man is home now. Everything will be as it was, and you will behave yourself. I know you get easily confused when stressed. I forgive you for acting a fool today with Leo and those criminals. I am even going to forgive this chastity mess, but you will have it back on the second we get home. You remember that I was loving and forgiving. I expect to have my Christian Axel grateful for my mercy. You did miss me, didn't you? You can say so and welcome me home now."

I nodded. "As you wish Master. Welcome home. I missed you." I lied like he told me to.

Master Jonas grinned then patted my head. "That is my boy. Come, let's go party with the other old farts." He took my leash and hand signaled for me to rise and follow.

Master Claus, Master Bladrick and Master Leo were already waiting for us at the stairwell. I dropped to a kneel behind Master Jonas when he stopped to greet everyone. He laughed with much humor while the Elders patted his back and welcomed him back.

Master Leo stood quietly stealing looks of concern at me. I noticed he focused on my near bashed in ear. I kept my gaze to the floor doing my best to fight back the urge to run to his embrace crying like a kid.

Master Jonas stopped the frivolity rapidly and we all took off down the back stairs. I watched all around me with anxiety. I knew this was stupid. I had killed Julius, but I couldn't shake the feeling I was being watched. Chills ran down my spine as the sight of Casper in his sunglasses filled my memory. He seemed so damned real, but I knew he was only drug induced hallucination. At least that is what I hoped.

We got into the Great Hall and all chattering halted as usual. The Elders puffed up their chests with pride. We all followed the attending black collar to the table. I followed behind Master Jonas, noticing the eyes of every Dominant in there was looking me up and down with what appeared to be lust. This was most unnerving. I thought I must be hallucinating again.

Master Jonas sat me between him and Master Claus. I was across the table from Master Leo but that was fine with me. Seeing him made me feel safer and somewhat calmer. The Elders shared wine and ordered a feast. Master Jonas beguiled them with tales of his travels. I sat there stealing glances at Master Leo wishing that I was laying in his lap at our haus listening to him read his silly romance novels.

He shot looks of adoration back at me. I could see the longing in his eyes. There was no doubt we both felt the same way. Our separation, even if only for three days, was painful. Master Claus decided that the atmosphere in the Hall was stale. He called the black collar attendant to play music through the speakers to lighten the place up.

Master Jonas chuckled, then threatened to get Master Claus drunk enough to do the dance of the seven veils for all of us. The crossdressing Elder had worn a gypsy type dress that seemed like many scarves attached together. The table erupted in laughter when in the background a familiar song began to play.

I looked up with a smile at Master Leo. He smiled back then stood up and came over to Master Jonas.

"May I borrow Mad Maxx for a dance, Jonas? My legs are getting stiff, and I want to show off my holding this beautiful boy to my old enemies Peter and Gretta." He lied.

Master Jonas had imbibed a few drinks and was feeling generous. "Sure thing, Leo. Just make sure you keep your hands above the waist. Not like you will have the choice, ja? You can show off to them rats all you like. Everyone at this table knows that you will never get the affection of our Priceless unless you tie him up. Ah, that was what today was about? You had to call in your buddies to try to fuck Maxx? Couldn't even subdue the boy without help."

He laughed out loud most cruelly. Master Bladrick and Master Claus covered their mouths chuckling and shooting humored looks at Master Leo. My Master just looked at the floor appearing upset by this unnecessary slight.

"Ja, okay Jonas, if you are done pointing out my failures with our Priceless. Can I get your permission or nein?" He kept his tone even.

Master Jonas nodded. "Go with Leo, Mad Maxx. Show this weakling the only moves he will ever get out of you, my love." I stood up without being told twice to take off with Master Leo.

Master Jonas reached out and slapped my backside as I went past him toward my Master. "Make him jealous of us real men, Mad Maxx," he shouted drunkenly.

I winced but took Master Leo's arm. I let him lead me far from the table. Holding him made me feel warm inside. He turned and faced me, taking me into his embrace. We allowed our song Nights in White Satin to sweep us away from all the horror of my shared leash with brutes like Jonas. My Master didn't have to say a word. I could hear him singing to me in my heart.

We were so enthralled with each other neither of us saw Gretta slip up to the Elders' table. She had come to tell Master Jonas she was withdrawing her demand he release her from purchasing my virginity rights. This surprised Master Jonas a great deal since she had been so adamant that he forgives her from her promise to secure her leash.

She told him that since she had been a private witness to my incredible skills with Master Leo, she could hardly wait to get her own taste of me. Master Jonas and the other two Elders were not only confused but stunned to hear that Master Leo had received my special services. Gretta told them she had never seen a more passion filled, loving thing in all her days. That my love for my Master Leo was not only obvious but beautiful to witness. The bonus of soiling his lover was too much for her to resist.

It must have been at this point she informed my jealous Master Jonas that not only did I provide Master Leo perfect service, but I had also reached my own equal one with him. Master Leo and I were oblivious to this foul bitch putting a wrench in the wheels of brotherly love between the easy to anger Elders. Turned out Master Jonas was not the only of the three to be pissed to find out I could find interest in a sexual encounter with a man, but not just any man, with the most disliked Elder Leo.

Normally Master Jonas would ignore rumors from the mouth of this weasel Gretta, but he could not deny the spark he was witnessing on the dance floor. Try as we might, there was no hiding our love when in each other's embrace. When the song ended Master Leo and I danced to a couple more then returned to the table. Gretta had already slinked off back into the crowd, her dirty deed done.

We noticed the mood had turned subdued immediately. I raised an eyebrow and Master Leo shrugged. He led me to my chair then turned to Master Jonas with a smile.

"Thank you for letting me borrow your collar, Jonas. I wonder if maybe I could come by later to return his things to him? He left his special medicine. I promise to be quick and then..." Master Jonas slammed his hand on the table then reached out and snatched my leash with suddenness interrupting Master Leo.

"Shut the fuck up, Leo. Christian Axel, you come with me right now. Leo, you stay away from me. You already have taken more than your fair share of my collar. I better not see your face, or I will rearrange it for good," he yelled while standing up.

He jerked me out of my chair with much fury. I fell to a kneel immediately trembling, unsure what was going on. Master Leo looked at the other Elders for an explanation for this odd threatening by Master Jonas.

Master Claus sighed. "Jonas calm down. Let Leo explain this before we all get angry and say things we don't mean." He didn't sound too interested in hearing whatever information he was requesting either.

Master Jonas bellowed out so loud all the chatter in the room stopped. "I don't give a fuck what he has to say. I am leaving. Leo, go die, you rat bastard." He took off dragging me behind him.

I practically had to run to keep up with him. I turned around looking to Master Leo with fear. He stood there appearing as confused at this scene as I was. I had no idea why the Vampire was angry, but I did know whatever it was, my ass was would pay for it.

I was right about that. Master Jonas practically kicked in his own door. He dragged me right to his bedroom. He knocked me to the floor then picked me up and tossed me into the bed. He pounced on me before I could crawl away.

I screamed and sobbed openly while he sexually assaulted me in the most brutal ways. When I tried to get away from the man, he grabbed my leash and held me still for his agonizing couplings. It was clear he intended for this to hurt as much as possible. He never said a word about why he was doing this to me. I was as terrified by his cruelty as I was tortured by his painful employment of it.

When at last he had his fill of punishing me, he pushed me from the bed. I landed on the floor nearly broken from the heinous session with this angered Master. "Go clean yourself up. You smell like fear and tears. I cannot stand the pussy in my bed when I thought I had a man. Don't you come back in here till you have the stink of it off you."

I nodded and grabbed my clothing that he had thrown all about the room during his attack. I limped in much pain into his bathroom. I closed the door still shuddering and sniffing back my tears of humiliation and despair. I went to the faucet and turned on the hot water. The steam began to rise as I sat there bleeding and battered from all my orifices. My chest ached with the agony of my pathetic existence.

I reached into the coats hidden pocket and calm came over me. I had the cure for my disease at my fingertips, the real one.

Chapter 50: Bonded to the Forbidden

The water steam boiled across the ceiling like a wet smoke. I shuddered and wept silently, caught in a trance of watching the cleaning stream barreling into the bottom of that tub. Within only a few moments the drenched air filled my lungs as I drew my breath. I knew the temperature for germ killing had been reached at last.

I took the sandpaper and that strange knife into my hands. I stood up and braced my flesh for the pain of the cure. I was about to step into the shower when the ground trembled around me.

"Halt, Mad Max, you bastard. Kneel, damn you," Der Hund's voice rung in my ears like the gong of a huge church bell.

I fell to my knees dropping my tools in my haste to mind my Master. The Core came through the wall hauling Maximillian behind him on the chain leash. I kept my eyes to the floor daring not to further anger Der Hund. I seemed to have been pissing off all my betters that day. I was in no mood to anger the most dangerous of them all.

Der Hund looked at the sandpaper and knife shaking his head. "This is not okay. Master Leo forbade it, Mad Max. He said no more of the scouring."

I whimpered in fear of Der Hund's words. "Master Jonas says we must clean away the foulness. I cannot mind his order without scouring. The illness is everywhere and deep too. Master Leo, didn't give us the cure water."

Der Hund frowned. "Ignoring one Master's directive to fulfil another is not the solution Mad Max. You and Mad Maxx are tired. This makes you all unable to see things clearly any longer. There is a way to fix this trouble without disobeying either Master, but your fatigue prevents you from acknowledging it. Master Leo finds the scouring marks unattractive. Master Jonas finds the soiled boy unattractive. Fine, Maximillian, you and your brother Christian shall partner and take the boy for the solution to this problem."

Maximillian nodded. "Your wish is my pleasure, Der Hund." The seductive shard come forward just as Christian appeared through the wall with an evil grin to join his brother.

I grabbed Mad Maxx and hauled Max with me jumping from the flesh. I was grateful to be getting the rest. We Max boys were beat. It was time to let the fresh shards take the wheel. None of us could take another moment of the horrors heaped upon the boy, not even our masochist was enjoying this torment anymore.

I shot a wicked, but tired, smile at Maximillian. "Everyone is trying to fuck the boy right up the tailpipe schwuler boy. You are just gonna love it, whore. Hope you brought your ChapStick and lube. You are gonna need both by the gallon."

Maximillian sneered. "You can go fuck yourself for a bit. I know you cannot get enough of that screwing yourself over. I have watched your dumbass moves, fool. Thank Gott the expert seducer is here at last. If you and the pain lover run the show another minute, this boy would be worm food. Let me show you how to calm down the Master before he tears the flesh to shit, idiots." He jumped into the boy before he could go catatonic.

Christian glared at me as Max, and I stumbled past him. We were pretty whipped, you know, it has been a rough few weeks. "See you soon pussy. Get that cane of yours from Master Claus. Maybe then you would be someone worthy of my respect again. Seems that fucking monster on your leash has made you weak. Keep him the fuck away from me schwuler freak." He pushed me and Max into the wall.

Der Hund reached out grabbing Christian just as I come at him fists flying. Max held me back by leash while the Core restrained Christian. A fight of epic proportions nearly broke out right there on the bathroom floor. Maximillian watched the show with a grin but didn't move the boy from his kneeling by the heated tub.

"Stop this shit, both of you. Christian, we discussed this. You have your assignment. Maximillian will aid you, but you will mind me or be shattered. I cannot have my two strongest shards attacking each other. Now get to the flesh and be still. Mad Max, you and Max go rest. I need you both for the aftermath of Maximillian's and Christian's mission completion. Do not anger me further. I grow weary of this existence. Any little thing will push me to give up my struggles. You all must serve me, or we all die. That is the bottom line. We work together or we are crushed as a team beneath the feet of the Dominants. Do any of you desire that we lose to these nothings?"

All of us shook our heads and said in unison. "Nein, your will be done." I took off with my beaten brother Mad Maxx and the well-worn Max for our deserved rest.

Christian shook off Der Hund. "I do as you ask my old friend. Time to put a few X's on our list, ja? Move over schwuler boy. The real man Christian is coming to back your pansy ass." He jumped into the boy.

Der Hund gave the hand signal for me and Christian to attend to our duties. We saw him slip back into the shadows without another sound.

I watched Christian take his place behind me at the wheel. I flashed him a look of suspiciousness. I didn't like having the killer to my back. I dared not complain to Der Hund about it though. I knew his presence was necessary for the completion of our mission. I decided to ignore him and focus on my job of cleaning up this horrible mess.

My wearing out caused the wrong shards to run the boy far too long. They were not designed to manage the schwuler sex and please the Masters in perfect service. I was most unhappy that they had managed to cause the flesh much damage, but I understood, they didn't possess the skills to prevent such abuses. I was here at last, ready to make everything better.

Christian lightly kicked me in my bottom with his foot. "Get to it pussy. What the fuck are we waiting for? This flesh is foul, bruised, and in bad need of a few weeks off from being a punching and poking bag."

I shot Christian a hateful look. "Der Hund said be still. I will get this illness cleansed. The rest we need isn't gonna happen. You let me do my job and I won't interfere with you when you do yours. Deal or nein," I growled at the lust/anger shard.

He nodded. "You got it whore. I have no interest in sucking cock nor playing the pincushion like the rest of you fatherfuckers. I intend to nap right here, until you get my targets where I can snap their necks that is."

I rolled my eyes at that. "You bet, Mr. Big man. Just remember to stay the fuck out of my way."

He chuckled but closed his eyes leaning back for his napping. I picked up my scattered clothing and killed the shower water. I grabbed the tools and dropped them back into the hidden pocket of my coat.

I admit I spent a moment looking at them with longing. It would be much easier if Master Leo had not given that stupid directive. What does everything have to be done the hard way for Maximillian? All I can say is bullshit.

With the stealth of a mouse, I opened the bathroom door. I listened in the darkness for the sounds of the Vampire's breathing. I could hear it was deep, slow, and rhythmic. He was definitely in a sound slumber. I had hoped he passed out from his many glasses of wine and his foul raping of the boy.

I winced from the horrid pain the second I tried to take a step. Yikes, the flesh was raw and unhappy about the recent rough treatment of it. I took a breath, bracing myself against the pain, then slipped from the bedroom.

In the hallway I let go my air with relief. I knew Master Jonas would be out for hours. He didn't get drunk often but when he did, he slept like the dead. I headed to his door and picked all the locks in only a minute. This Master, unlike Master Leo, didn't go that extra mile to keep me from getting out. Escaping his apartment was so easy, even a novice at picking could do it.

I went into the hallway with silence closing his door behind me. Then fast as my feet could go, I ran for the back stairwell. I hit those steps, nearly falling, I was moving so quickly. I tore down them but kept my eyes behind me. I knew Christian killed that rapist, but Mad Max told me there could be another one of these monsters. I took that seriously.

I would stop on the stairs ever so often and listen to make sure I heard no other footsteps coming.

I threw myself up on the wall and peeked around every corner when I came to them. I felt like an undercover agent trying to keep from being discovered by the enemy. It was enjoyable in a spooky kind of way. I confess, Christian and I did have a bit of fun with it.

When we got to the main floor, he and I would hurl the boy behind walls and pillars. Even when the person coming our way was nothing more than a silver or black collar. He and I made a bet we could make it all the way through the Haus without a single set of eyes spotting us.

It seems very silly to me now, but I even hummed the music from that show Mission Impossible. I saw a couple of the episodes of that series on one of the Master's TV a few times. I thought that Dan Briggs was very cool and wanted to be as smooth as he seemed to be. Back then, I wasn't aware he was only some actor. No one ever told me he wasn't a real person.

Anyway, the behavior worked. We were not noticed sneaking to the pool room that Master Leo had taken us to. When we arrived there, I stood at the entry peering inside to make sure everyone had left the swimming for the day. I was in luck to find the pool void of people.

I walked into the huge place looking all around it in awe. When Master Leo brought me, he didn't allow me to investigate all of the stuff there. He was only interested in the special services, you know. I was kept tight on his leash. This time, I could examine anything and everything I wanted without being told nein. It was amazing.

There were chairs made of plastic stuff for relaxing. I marveled at the tables with all kinds of magazines to look at under each one. Some were of cars, others had pretty girls wearing almost nothing. I stopped and leafed through a few of them. I smiled with much joy to find the ladies in them wore only the swim panties and bras.

Christian and I both agreed that before I left, I needed to grab at least one or two of the sexy girl mags. I knew I would like to see more of those girls the second I had some alone time. For a change, Christian and I both thought this a clever idea, no quarrels about it.

When I was at last able to tear my lustful eyes from those beautiful females, I went to the edge of that pool. The water glowed with bright lights under the surface. It was so blue it rivaled the sky. I sniffed the air and smelled the bitter cleanness of chlorine. Just being near that magical stuff made me feel better.

I stripped from my clothing to nakedness barely able to control my urges to jump in. I managed to get my last sock off before leaping into the shallow water. The warm, inviting liquid caressed the flesh like a desired lover. I closed my eyes and dropped down to immerse myself in the mystical cure for my diseased flesh. I stayed below the surface until I could hold my breath no longer. I popped up laughing with glee at this wonderful sensation of being clean at last.

I splashed, dropped, kicked, and giggled having a grand old time for quite a while in the pool. I played a game of tag with Christian till both of us tired of being trapped in only half the space available. I walked to the marker that indicated the water was too deep for me to stand many times. I looked at the forbidden side with longing.

Christian wanted to know what it was like over there in the deep too. He asked me if maybe we could swim if we tried. How hard could it be? I practiced the way I had seen on the TV in the shallow end for a bit. I often sunk to the bottom, but I was sure once in the larger area I could keep that from happening. I went to the drop off once more, took a breath and stepped off.

Almost immediately I began to drown. Panic filled me when my feet could not find the bottom, and my head would not reach the surface. My arms and legs flailed uselessly as I sunk deeper into the pool.

In only minutes, I could not hold my breath any longer. I could not reach the air, but I didn't have the gills either. I realized I had no choice but to suck the water into my lungs. When that happened, I knew I was finished. I closed my eyes and prepared for the end.

Suddenly I heard a splashing noise. Strong arms pulled me from the drink right to the surface. I took a loud breath thanking Gott that I could get air again. I was so scared I barely noticed I was being pulled along by someone swimming next to me. All I cared about was that I wasn't going to die.

The swimmer hauled me back to the shallow end and let me go. I rushed to the side of the pool grabbing the lip of it for dear life. I was trembling in terror at that near miss with the reaper. That was indeed a close call. I turned to see who my hero was and near pissed the pool at the sight.

Master Malfred stood there in his undershirt and pants panting with paleness to his skin. He was watching me with a look of fear in his expression. I shot a look at the pile of clothes strewn in every direction from the entry of that pool room to the deep end. He must have come in to find me drowning. It appeared that in a panic he ripped off most his clothes then jumped in to save me.

I shivered in fright realizing this voting Council member was one that I suspected was the rapist, well before Julius admitted to it. Even if he had been cleared of that crime, he was an associate of Peter's. Master Jonas had told me he also had been demanding a weekly leash with me over the loss of his expensive metal Tamina.

This was not a good thing, being alone with a man I knew was looking to taste the Priceless. He stood there still catching his breath, saying nothing. It was beyond unnerving. I decided it best to tell him I appreciated his aid; then get the fuck out of the pool fast.

I coughed out the water I swallowed for a moment while still trembling. "Thank you for saving me, Master Malfred." I began to slowly move toward the steps.

He nodded. "Mad Max, where the fuck is Leo? Or Jonas? Or hell any of your stupid Masters? They just let you wander the halls off leash at night?"

I shook my head. "Uhm, nein? I uhm, came here for a quick swim while Master Jonas naps. He was tired and I was not. I will need to get back to his apartment. He should be awake now. Thank you again for the mercy Master. I apologize that I was a bother." I moved closer to the steps ready to run naked from that room if necessary.

Master Malfred scoffed. "I don't believe a word of your story boy. You snuck down here and none of your Masters are aware of how close they just came to losing their collar. This is the second time in one fucking they I find you neglected. I see the marks of rough treatment all over your flesh as well. They abuse and ignore such a treasure? What the fuck is wrong with these idiots. I would think at their ages they would be more careful with their playthings. Maybe that is why the Elders are never supposed to hold the silver. A Priceless is too valuable to just toss around without care. You cannot be replaced if they break you."

I shook my head and dropped my gaze taking another few steps toward my escape. "Oh, they are good to me Master I swear it. I must be going now. I will let them know you helped me. I am sure that a reward is due." I took off for the steps fast as the water would allow for my movements.

"You halt right there, Mad Maxx. You know better. I am one of the Voting Council. You will show the respect due me. I didn't release you from my hold, did I? I think not. You will kneel right now. This water is shallow enough for the proper protocol. Refuse me and I will to your Masters and demand punishment for it," he yelled out just as I made it to the steps.

I winced then knelt immediately before my escape from this bad situation. I knew Master Malfred had me dead to rights. If I dared to show insolence to one of the voting Council, hell any Dominant for that matter, he would have the right to have me whipped to death.

At the very least he could tell Master Jonas I slipped out. The boy simply couldn't handle another cruel raping from the Vampire for at least a couple days. I decided my only real option was to try to talk my way out of trouble with this scary man. I had to find some way to keep him from talking to Elders as well.

He walked over still grumbling about my disrespectful behavior under his breath. I shook in terror as he reached me and stood there staring, with an expression of irritation on his face. He appeared to be looking me over as I knelt there in silence neck deep in the water.

I stole a good look at him. I had never bothered to notice much about the man until that moment. Master Malfred was almost six foot four. He was clean shaven with a chiseled jawline and deep-set hazel eyes. His hair was chestnut brown and full of waves. This Dominant was known for his constant working in the gyms. That had resulted in him obtaining muscular build with much strength that he hid under expensive suits.

Despite his constant effort to maintain this strict regiment, he strangely was not known for showing off his bodybuilder physique. I recalled that girls in the washroom often would speak of this Dominant in wishful statements that they would catch his eye. Annette and Geraldine referred to him as dreamy, handsome, and a hunk.

Master Malfred was also known for his highbrow" taste in everything, including his silver. Master Malfred would only wear the finest clothes, or lay claim to the top-level collars. To be purchased by this Dominant was the dream of most of the pleasure submissives at that time, both the males and the females. If he ever held their collar, that marked them as being judged the best.

Tamina was a good example of his very caution choice in the submissive he chooses to be his own. She had been renown for many years as not only expensive metal, but the most beautiful of all the females.

Like many of the Dominants of the Haus, Master Malfred was the bisexual, not the pansexual as this guy was super picky. All his past male silvers had been of the highest levels obtainable and of only the most perfect of builds. He was meticulous, cultured, and expected all his pleasure submissives to be the most desired of any in that Haus at any given time.

When the silver collar was leveled by age, accident, disease, or just lost their status he would trade, sell, or auction them off to be used by another. Well. all them before his precious Tamina that is. She was the jewel in his crown, having the distinction of being that lucky silver that had it all.

The woman was rumored to be talented in the bedroom, artistic in dance, could sing like the canary, was gorgeous, and had the temperament of the saint. It was never any secret that Master Malfred was desperately in love with this girl beyond all reason.

He even tried to make children with her. When after many years the aging Tamina, she was thirty-four at this time, and Master Malfred, he was forty-one, separated, it had been discovered this Dominant was in fact infertile. He blamed a late age case of the mumps for his lack of life-granting seed ability.

Her pregnancy by Olaf must have been quite the blow to this Dominant's overblown ego. It also apparently had broken his heart. Rather than marry her or even blood bond, he had sold her away and all his other silvers went with her. It was said around the hallways of the Haus that Master Malfred determined if he could not hold the finest of all the collars, he would hold nothing at all. He believed he deserved only the best and that little bird was no longer available.

I didn't know a thing about if this Master Malfred was the hunk, or a snobby blue blood of old money. All I cared was that this huge man was a friend of my father, and he seemed to keep showing up at all the most inconvenient times. *Okay, yeah to be fair, he kept me from drowning but what the fuck did he want from me? He only did what anyone should do if someone is in trouble. I wasn't willing to give this fucker a blow job for being decent. Shit. I wished he would move on and let me be.*

Master Malfred sighed loudly then finally spoke. "You know you have me in a bind Mad Maxx. If I just let this go, then maybe tomorrow night I don't come in for a nightly swim in time to save you. If I report your sneaking down here, then these fools that hold your silver will savagely beat you. You already are covered in enough scars as it is. They're going to ruin a beautiful boy with their careless cutting and beatings. Tell me what the hell am I supposed to do with you?" He crossed his big arms appearing perplexed.

I shrugged. "You can just let me go. I will swear to never come back here. Master, you should know I am a man of my word. There is no need to upset my Elder Masters. This will not happen again. I learned my lesson."

He snorted. "I am sure you are a man of your word but you my dear are the Priceless. That means you can never be sure of what you may do. That madness of the forbidden silver clouds your good judgement. Well, I cannot sit back and just let this problem slide. after what I saw today, I have been made a believer that you are indeed worth the metal you wear. Your silver is simply too precious to risk. The answer is clear to me. I have no choice but to teach you to swim."

I looked up with a startle. "Huh? You are going to teach me to swim, Master? Why would you do that?"

He smiled at me, appearing kindly. "Well, if I tell on you, they will scar you up. If I walk away, next time maybe you will drown. If I teach you to swim then there won't be another incident like this one I come across tonight, ja?"

I stared at him with suspiciousness suddenly recalling Mad Max had bitten the ear of his attacker. I narrowed me eyes and looked at both Master Malfred's ears seeking damage. I witnessed no marks, bruises, or redness. This was not Casper the bank robber. Mad Max must have hallucinated the second rape, or it had to be someone else. That thought made me shudder for a moment.

I shook my head. "I am not worthy of such a gift from a Dominant of your level Master. I would beg you to obtain a black collar for such a low task. Though I do thank you for the mercy of your not reporting me to the Elders over this mistake."

He chuckled deeply with brightening eyes. "Nein, I would never allow the disgrace of a black brute messing around with the Priceless. My dear you have no idea how much that silver of yours is desired around this Haus. You are a legend in your own time, and everyone wants to have a bite of the mythical Mad Maxx. Such a temptation is too much for those unaccustomed to the finest the world has to offer. This is settled. I will teach you to swim or we can go right this minute and awaken that uncivilized brute Jonas from his abusive, drunken stupor, I mean peaceful nap. Up to you Maxx. I am the expert at swimming you may have noticed."

I groaned. "You seem to know a great deal about many things Master Malfred. I assume you learned much from Master Peter."

Master Malfred shook his head/ "Peter? What the hell can that man teach Malfred? Nothing. The man is a moron. He throws away a fucking Priceless and holds dear the fickle Agnette while teasing the brutal brute Grisham. He has no class nor eye for quality, which is for damned sure. My Tamina never did like him, you know."

I nodded. "As you say Master. I am not in a position to argue with you about any dealing you have among my betters. I will take your deal if you swear to never breathe a word of my being here tonight to my Masters or anyone."

He grinned with thrill. "That you can count on. I give you, my word. Now, we have the lesson, ja? I am going to require dress service, Mad Maxx. My clothes are soaked or scattered. You will provide this service for me then we return to the water and teach you to float." He took off for the steps stopping only to grab my leash.

I followed him to one of the tables. He stood still to allow me to remove his wet jogging pants, boxers, socks, and undershirt. Master Malfred had me lay them over the pool chairs to help them dry out quickly. He then had me fetch his tennis shoes and workout t-shirt. This was not his usual Italian suit and fancy leather shoes. I realized he must have come to the pool after leaving his gym work out.

I noticed he was watching me closely as I provided the dressing, or undressing in this case, service. He didn't say anything but tended to use hand signals rather than speaking his wishes. I didn't hesitate or forget any of my graceful skills which appeared to please this picky Dominant a great deal. When I finished all that, he asked I dropped to a kneel at his feet waiting for the next instructions.

Master Malfred chuckled. "Well, I take back one thing I said about Peter. He is the baboon in his own behaviors, but he is most skilled at training a submissive to perfection. I have not been so well pampered since my Tamina left my collar. Bravo, Mad Maxx. I am impressed. Not many a young man can show such grace and pose. You are indeed top-quality silver as I suspected. I find Peter even more ignorant than I ever imagined. He threw a diamond out the door and thieves found it. I see them all fighting to possess it all alone, and I don't blame them. They are bigger fools than Peter. I still have trouble believing they let you wander off alone without protection of any kind. That is lucky for Malfred, because I can pretend for this moment the Priceless is all my own. Ha, enough of this silly wishful thinking, ja? Let's go for that lesson. I am eager to see how fast you can learn something new, meine Taube (dove)." He took up my leash and hand signaled for me to follow.

We got into the water and Master Malfred had me hold still while he took me in his arms. I panicked at first till I realized he was holding me on the surface of the water to practice my swimming. He gave instructions on how to turn my head for breathing, circle my arms and kick my legs. He stated if I could do this, I could do a simple breaststroke.

He let me go and I managed to get across the shallow end of the pool a few passes using this technique. Master Malfred then had me cease and he grabbed me again this time holding me face up. He told me to hold my breath and relax. Then slowly he took his arms from under me while I took in air held it then let it go and rapidly took another deep one. Within only a few minutes, I had learned to float.

I admit, I was starting to relax and have an enjoyable time. Master Malfred was patient with me and never offered a negative word when it took me a few tries to gain a skill. He would smile while saying many compliments every time I obtained another ability to navigate the water safely.

I was far from being a strong swimmer but in only a few hours he had taught me enough to prevent my drowning. I was feeling unhappy that soon I would have to leave and sneak back into Master Jonas's apartment. I had almost forgotten his brutal rape; almost but not quite. I could have been happy swimming with my new friend Master Malfred for the rest of the night rather than return to that hateful Vampire.

Master Malfred told me that he wanted to be sure I could survive the deep end if he was not around to save me next time. He swam across the pool then used his hand gesture telling me to come to him. I took a deep breath to steady my nerves, I was scared after what happened earlier you know, and followed his command. I used my simple breaststroke to cut through the surface and arrived safely to be greeted by the smiling Dominant.

He held on to the side instructing me to do the same. "Magnificent. You are a real gem. Intelligent, strong, graceful, polite, resourceful, and beautiful to boot. I will not lie. I am more than a little jealous of your Masters. Lucky bastards indeed. If you had been any silver other than the Priceless one, I would be tempted to steal you from under them. Nevertheless, you need to return where you belong meine Taube. There is no reason for me to torment myself any further. You have learned to swim, and I have no doubt in time you will be an expert at this skill. Come, I race you to the shallow end. You can give me the dressing service, dress yourself, then await my lead. I will take you back where you belong. No arguments. This Haus is not safe for your traveling unaccompanied by a Dominant. I better never see it happen again or next time, I will have to report you." He looked at me with sternness in his expression.

I nodded. "Thank you for the mercy, and the lessons. I will not forget your kindness this night. It is not often I am treated so fairly."

He frowned. "Ja, I can see that by the marks on your face and shoulders. These handprints, do the Elders find you struggle in the granting of special services so fiercely they must hold you down and beat you?" He pointed at the marks Master Jonas left from his rape and smacks to my face.

I dropped my gaze feeling uncomfortable all the sudden. "It is not my place to discuss such things, Master. I apologize for the refusal to answer your question, but you must forgive me for keeping my pleasure submissive oaths. The details of special services to my Masters are forbidden information to share."

Master Malfred nodded. "Ah, ja, you are indeed worth your metal meine taube. Well versed in your protocol, and perfect in your loyalty to your Dominants. You take my breath away. Such a refreshing thing to behold in this Haus of poor quality and common silver. Enough of this chit chat. I race you to the other side. Eines, zwei, drei, gehen." Master Malfred took off swimming at full speed.

I let go of the side of the pool and chased after the swift moving Dominant. He of course beat me by miles, but I did manage to make it with an adequate stroke to keep me from drowning. He laughed and ruffled my hair when I reached the steps panting and shuddering from the racing.

"You need to work out, little one. A few days a week in the gym would do you a lot of good. You're a strapping young man, but not as big as you could be. You better get those muscles fed up to girth while you enjoy the best years of health, ja?" He took up my leash and led me out of the pool.

I narrowed my eyes. "Pardon me for my ignorance Master, but is such a thing permitted?" I knew the Haus had a workout room, but I thought that silvers were not allowed dare.

He pointed at a row of lockers with a nod. "Go get a towel from the shelves. Ja, you can go to the gym long as a Dominant of high status is with you. Your Masters never told you this? well shame on them. You ask them if one will grant you this mercy and if they say they are too busy, come tell me. I will be happy to take you in their stead, but you must get their permission for that. I want no part of pissing off the grouchy Elders. My being seen with you too commonly gets the tongues wagging you know."

I retrieved his towel and began drying him off in prep for dressing service. "Ja, I can see why you would worry Master. My Masters do not care for you very much, but that you know. I apologize for their tendency to hold grudges. That is their pleasure, and I am their property. I thank you for sharing that information and for your generous offer. It will not be necessary though. I have a Master that will be happy to take me if I ask permission." I smiled with brightness as I thought of Master Leo. Leo really is the greatest. Maximillian winked at me, making me giggle.

Master Malfred smiled. "Well, that is wonderful, meine taube. I will be seeing you at the gym then. I can even give you a few pointers to build you up faster. When you become the Dominant, come see me and I will test your strength in a wrestle, ja." He ruffled my hair again while chuckling loudly.

I paused my dressing him with a startle. "You heard I will break my collar, Master? Who told you this?"

He shook his head. "Mad Maxx don't be silly. There has never been a Priceless in all the history of this place not seeking such a pleasure. This time, it will not be just an unrequited desire though. You meine taube are a legend, and the only one I ever believed capable of such a magnificent feat. You have survived Xavier and all his hideous men. You even managed to obtain the thrill of the entire Elder floor. I for one will vote ja in your collar selection. A spirit as big as yours is too big to be held to the floor by the weight of silver. You can count on Malfred to see you set free."

I nearly squealed in joy at this most unexpected, good news. "For truth you will back my bid for Dominant, Master? I don't know how to thank you enough for such a mercy." I grinned so large it hurt my face.

Master Malfred laughed loudly. "You will find a way no doubt, Mad Maxx."

I suddenly felt a chill run down my spine as I recalled the Voting Council had a tendency to abuse their positions of power during collar selection. I took a deep breath and looked at the floor.

"I think I am beginning to understand you, Master. I apologize for my ignorance of your attempts to woo me. Is my compliance with your request for special services what you are seeking?" I felt the fool for falling for this helpful friend act the powerful Dominant had been playing on me.

His smile faded and he looked upset. "Nein, I would never force such a thing for my favor with the vote. I am not so easy to buy Mad Maxx, nor should you be willing to sell. I should be offended, but based on the marks you carry, enforced treatment is all you know. My lovers either are cared for as my submissive or come to me willingly without expectation of re-payment or I go without. I need not beg nor trick anyone for their affection. You take a good look at me Mad Maxx. I am the desire of every eye, and you meine taube are too. Only the best is worthy of our attention. You would do well to learn that lesson."

I winced at his words. "I beg your forgiveness for my misunderstanding, Master. Thank you for the correction of my errors. I had no right to question the motives of your kindness towards me. I can do nothing but ask your punishment for such a foul insult." I felt the idiot realizing I may have just blown my friendship with this most necessary Dominant.

Master Malfred shook his head still frowning. "Nein. Punishment denied. I told you I understand you knew no better. Now you do. Don't let that happen again. I offer you a lesson in swimming and you gave me dressing services in return. We are even. I give you my vote of ja for the Dominant selection because you earned my respect. Beware not to question my good manners or you will lose it. Finish this task with haste. The hour grows late, and you need to get home before that ungrateful Jonas awakens. I would like to see you again to swim, but these marks hurt my eyes. Next time I see you, I would like to enjoy the view of your flesh unhampered by such cruelty." He held out his arms awaiting me to put on his shirt.

I finished dressing him, then moved quickly to redress myself. He sat in one of the pool chairs watching me in my task. I was feeling very sheepish about my assuming his designs were of the shady kind. I tried hard to console myself with the knowledge that his playing fair was the exception to the rules of the Haus, not the usual. How the fuck was I supposed to know, ja?

When I was fully dressed, he took up my leash and stood there staring at me. I kept my eyes to the floor wondering if he was going to berate me further for my stupidity. Then he cleared his throat and looked at the ceiling.

"I see I have frightened you, meine taube. That was not my intention. I am ashamed that I reacted so impulsively to your picking up on my interest in you as a lover. You merely followed your instincts and asked me outright what my motives were in aiding you this night. Well, I will admit, I am desirous of a relationship of the intimate kind with you. That said, I will not dare to make such a pleasure contingent on my vote for your collar selection. If you choose to be my partner, then it would be for the joy of such a coupling only. If you say nein then we remain friends anyway. It doesn't change my respect nor want for you." He looked at his shoes awaiting my response to his offer for me to become his boyfriend with benefits.

I confess, I panicked at his request. "I am most flattered you would think of me for this role in your life, but Master I must decline. I wear the collar of four others. My special services are not my own to gift." I breathed out slowly trying not to hyperventilate in terror of saying nein to a man that could trap me in my metal for all time.

To my surprise he chuckled, then ran his hand down my check lovingly. "Ah, you do wear the collar of four Masters. However, I know because you do, they could not order monogamy in anything other than that penetration virginity of yours. I am the top not the bottom. I happen to know you are the penetrated by default. Your special services are yours to grant to anyone lucky enough to hold your favor."

I shook my head. "This you say is truth, I will not lie Master. That said, I have no room in my busy schedule for a lover of either gender. As it is, my preference is for the female. I do not seek a lover of your qualifications, but as I said I am most flattered you think of me this way. I endure what I must for the pleasure of those that own my silver. Other than that, I would rather die than suffer yet another one sided, unappreciated, and painful intercourse situation that is not necessary for my survival." I closed my eyes and swallowed, biting back the rising fear that Master Malfred would hit me for sure for refusing him and admitting I hated sex with the man.

The blow never came. "Fair enough Mad Maxx. It is too bad for Malfred that you don't find me of interest. However, I happen to know you do have the capacity to enjoy the act of a couple with the man that is not survival oriented. Meine taube, Gretta saw Leo and you in this very pool a few days ago. That big, mouthed harpy has told everyone in this Haus of the incredibly skill and beautiful special services granted your Master Leo. I wish you will reconsider my offer. I can do more for you than Leo ever can, and I assure you I can love you deeper. Or would you call Gretta the liar?" He smiled with humor and tapped my cheek in gesture for me to open my eyes to look at him.

I did as commanded feeling my heart speed to amazing rate. "She saw that. Oh shit, which is what Master Jonas meant at the Great Hall. Why he was angered. Oh no, Master, please I must go right this minute. If Master Jonas awakens and finds me missing, there won't be anything left of Mad Maxx for you to inquire about." I fell to a kneel, trembling in absolute blind terror.

Master Jonas knew about Master Leo and me. I was toast. The jealous Vampire had been trying to get me to find lust with him for more than a year. If Gretta did see that private scene, she would have witnessed my sexual interest in Leo's touching. Shit. No wonder Master Jonas blew his top. I would have to work hard to calm this boiling tea kettle before it caused a deep rift in the Elder brotherhood. Worse, Master Jonas no doubt would be looking to take Master Leo's head from his shoulders, or at the very least the one below his waist.

Master Malfred appeared shocked by my severe reaction to his reporting the gossip to me. "Uhm, okay sure Mad Maxx. I am happy to take you home. What of my request? Is nein the definitive answer?" He flashed a nervous smile.

I looked up at him with an expression of frustration that I could not hide. "Can I have a few days to consider your generous offer, Master? I need to think about such a sincere relationship." I wanted to buy a little time to get him off my tail and calm the Vampire as I was under too much stress.

He nodded and blew out his breath as if relieved. "Of course, meine taube. You take your time. Think it over carefully. I want you to come to me willingly or not at all. Now, let's go. Enough of this banter, ja?" He took off releasing me from my kneeling.

I wanted to sneak over to Master Leo's apartment to warn him, but Master Malfred insisted on walking me all the way to Master Jonas's place. I noticed he was careful to keep anyone from the sparse night shift staff from seeing us.

I was grateful that he understood, without me saying so, that seeing the two of us together after the other embarrassing rumor floating around, would be devastating to us both. I was very aware of Master Jonas's feelings towards this man, and Master Leo liked him even less. I simply didn't need any more trouble with my jealous Masters.

We made it to Master Jonas's door unviewed and safely. Master Malfred looked nervously down the Elders hallway appearing most uncomfortable. He let go of my leash and ruffled my hair with a smile, after being sure we were alone.

He leaned in then whispered in my ear, "I had a lot of fun, Mad Maxx. No matter what you choose regarding my inquiry, I would ask that we do the swimming together again sometime soon. I haven't felt this good since losing my Tamina. Thank you for putting the spring back in my step and putting a smile back on my face. Now you remember what I told you. Do not wander this Haus alone anymore. There are those that will not take nein for an answer. I fear for you with so many looking for a taste of your metal. You get inside there with that lucky bastard before you get yourself in difficulty."

I nodded. "Thank you again for all you have done for me this night, Master. I will let you know my answer soon. I wish more Dominants in this Haus were such gentlemen as you are."

He smiled brightly then with much stealth he slipped into the darkened hallway and sped off down the stairs. I opened the door and slipped into Master Jonas's apartment holding my breath. I locked all his locks and stood quietly listening to his slumber snores. I almost fell to my ass in relief when at last his buzzsaw sound echoed down the hallway. I had managed to slip out and get back without discovery. That was the only thing in my favor for the moment.

I paced the living room for a bit wringing my hands in panic. I needed a plan to answer for that private moment Gretta was telling everyone about. I thought of everything from temporary madness to calling the woman a liar outright.

In the end I decided saying nothing at all was the best route. Master Jonas would never hear another rumor. Of that, I would make sure of. Master Leo and I would have to keep all our trysts in our home, and no more slipping out into public. I knew that in the Haus rumors were like leaves on the trees, no one showed interest for long in any single one, there were so many to keep one's attention.

I went to Master Jonas's bedroom at dawn and removed my jacket and boots. I sat down next to his bed and closed my eyes for a quick nap. I was still very sore so sleeping was not coming with any depth. It seemed only minutes before the Vampire awoke with a foul temper still reigning in his heart.

"Well? You sit there on the floor and deny me my cuddling? I leave for two weeks, and it seems you sided with the enemy, ja?" I opened my eyes to see the Vampire glaring at me from his mattress above me.

I shook my head. "Nein Master. I do only what is demanded of me by any of my Masters. That is my job."

He laughed with bitterness. "Oh? Well, that is not what I heard. My Christian Axel the straight boy found an orgasm with the pussy Leo. Interestingly, he has never even gained a true moan of passion during his intercourse with his man. How can this be? I would think you are in love with Leo and hate your own blood bonded lover, Jonas."

I looked at the floor with a sudden flashing idea that could pull my ass out of the fire. "That is the problem, Master. Master Leo is the female; he reminds me of the Frau. For a moment there I was able to fool myself but understand this only happened the one time. I was heavily medicated you know, plus the excitement of the pool perhaps. That was all it was. Master Leo was insistent in his services, and I am tired of his nagging. He has the right to special service no matter how much I dislike that fact. I don't enjoy the interests of Master Claus or Master Bladrick neither, but I do my fucking job as I was trained to do. Master Leo is the gossip, girlish, frilly type. Easy to mistake you know. You meine Man, are the epitome of the male. I could be drunk as the sailor and never mistake you for the fairer sex." I held my breath hoping this explanation would fly.

The Vampire growled and groaned. "You do have a point. Leo is the fucking girly man. However, you never find lust with the crossdressing Claus."

I chuckled with much humor. "Master Claus wears a dress but has more facial hair then Mistress Cora."

That made him laugh too. "Ah, you are right there. I still am not so sure you are being truthful about Leo. I saw the way you looked at him dancing last night Christian Axel. There seemed to be a lot of adoration in those baby blues of yours for that schwuler nelly."

I shook my head. "Master, forgive me for saying this, but you had a lot of wine. I think you heard this foul thing said and saw what you been told to see. I heard this rumor like you did and ignored it. I thought you wiser than to buy the bullshit from that source. Had I known you would be injured by it I would have warned you. I beg your mercy for not protecting your heart better. I will add that you are my Man, and my feeling for you have never changed."

Well, I wasn't lying there. I hated him just as much as I had from the start. His bondage enforced blood bonding me without offering choice had never been forgiven. I will never ever get over that trick.

Master Jonas paused a moment as if deep in thought then looked at me with a softening expression. "Okay, I admit that I got a little drunk last night. My memory is full of the demon alcohol's fog of deception. I have decided to take your answer for truth since you have always been honest with your Man in the past, even to the point of agony. I better never catch you engaging in anything other than what is required by that fucking rat Leo though. I am currently researching a way to end the hold the others have over my collar anyway. This will soon be of no consequence to me or to you." He smiled, then ran his fingers through my hair. *What the hell was with all the interest in my hair by the way. Weirdos.*

I looked up at him with a startle. "Huh? You say you found a way to snatch my collar out of the hands of the Elders to be exclusively your own? How?" Fear gripped my heart at the thought of losing the sanctuary Master Leo's home provided me from this brute.

He grinned with his toothy smile. "Ah, I have been working on this issue for months now, and finally I have the answer to all our woes. I am your blood bonded. I am aware there are little-known provisions for one with my special attachment to you. Even Claus seems to be unaware that as your only man, I can refuse to allow them any rights to the special services. That is all any of them really want of my Priceless. Well, I will demand the Council honor my complete control of who fucks my collar. I assure you the second Claus, Bladrick and that sonofabitch Leo, cannot get their hardon relieved by you, they will lose interest in holding your leash. I intend to present my case before the head of the Voting Council the end of this week. By next Monday, you meine heart, will be home for good with your man. No more of this sharing shit."

I trembled at his cryptic words. "Oh, ha-ha, very clever Master. I guess then I better collect all my things scattered in the other apartments. Perhaps I could go now to do this?" I was in a full-on panic.

Master Jonas chuckled with much humor then grabbed my upper arm hauling me roughly up into his bed. "Sure thing, Christian Axel, right after I spend a little time calling in my manly rights with you." The Vampire pulled the flesh into a straddle on his lap while kissing me with much wanton roughness.

I did my best to appear interested in his lustful actions, but not only was I not ready for anymore special services with anyone, but I also needed to immediately see Master Leo. I was not aware how many blood bonds were permitted per submissive, but I did recall there was more than one.

I was not happy to have to request my Master Leo to engage me in that agonizing ritual of Haus approved marriage with their submissive. However, there simply was no other way to prevent Master Jonas from keeping us apart.

I had to move quickly to make sure the blood bonding with Leo was done, witnessed, and recorded before the case went before that hateful Gretta on Monday. I was more than a little aware Gretta had every reason to find in Master Jonas's favor regarding his complaint. Shit, this was bad. Though I was more than happy to be free of the London Bridge, so there was at least a little good news.

I decided no matter how worked over the boy was, if Master Jonas wanted his special services I would comply with feigned eagerness. I hoped this acting job would calm him enough to obtain a release to visit with Master Leo.

I kept my fingers crossed this sacrifice would be worth the pain as the Vampire's groping got more heated. It took all I had to hold back my fearful groaning when I felt his manhood began to rise under me.

He began to paw at my blouse trying to remove it when a loud knocking broke the sounds of his lustful panting. We both held still listening as once more the sound of pounding echoed down the hallway.

Master Jonas sat up nearly knocking me to the floor off his lap. "Who the fuck is bothering me at this early hour?"

I shrugged. "I am not expecting anyone, Master. Shall I answer the door?" I silently thanked Gott for the interruption.

He nodded. "Ja, go tell them to go the fuck away. I wish for no visitors for at least an hour. I have services owed me." He slapped my backside with a hearty laugh as I jumped free of him running to find out the identity of my savior.

I opened the door to see the grouchy faces of Master Claus and Master Bladrick. I dropped to a kneel immediately as they walked past me not bothering to wait for the invite. They took a seat on the couch together still glaring at me.

Master Claus coughed then growled out. "Where the fuck is Jonas? Get the hell up and go tell the man we are here for our leash rights, Maxx." I looked up at them with a startle.

"Leash rights? I don't understand Master. I thought this was Master Jonas's cycle?" I was beyond shocked at this news.

Master Bladrick scoffed. "If Jonas had bothered to look these two days belong to Claus and I. Leo has your leash for three days, then our two then his three. We were most generous to allow him the night to re-bond with his man, but now we are here to collect what belongs to us. Go get your things and call Jonas in here, Gott damn it. Make us repeat ourselves and you will get a thudding." I nodded then rushed from the room to follow my Masters command.

Master Jonas was stripped and waiting in the bed for my return. I walked in and dropped to a kneel next to the bed.

He growled while sitting up and reaching out to grab me. "My cock is not down their Christian Axel. Quit playing games with me. Get the fuck into this bed and get to servicing your Master."

I dodged his eager hands. "Master, I beg of you I must deliver a message. Master Claus and Master Bladrick are in the living room. They claim this is their clock and not yours. They demand I leave with them immediately or risk the chains for it." I never thought in a million years I would be happy to leave with those two, but that day I was. Anyone but that brutal Vampire was looking mighty good to me at this point.

Master Jonas sat up with fury in his dark eyes. "What? Fuck this sit. Get my robe. We will see about this." I did as commanded and followed the storming Master down his hallway.

I stayed in a silent kneel while the three of them argued. No matter what Master Jonas tried he knew he was beaten. The Elder Masters were correct. The days did belong to them. He eventually relented his arguing and gave up. He snatched up my leash and began to drag me back to his bedroom claiming I needed to get packed.

He turned around for a moment and yelled, "Enjoy it while you can, you old farts. This will be the last time you come into my Haus and haul off my man to befouled with your lust. I am sick of this shit. Mark my words, your time is finished." He then turned and dragged me back down the hallway.

He forced me to the wall by his dresser. Then with much anger ripped down my breeches to my knees. He told me to stand still and keep my mouth shut. I watched him dig through his drawers until he retrieved another metal chastity device. I shivered in fear as he put it on my manhood and hodensack then locked it in place with a small padlock. The hex wrench would no longer be of use for this cock cage.

The Vampire smiled with delight as he admired his cleverness at restraining my lust. "Now let's see you find your pleasures with Leo or anyone, including yourself. Now, get your things and I will see you in two days. You better hope those old stallions go easy on you boy, because Jonas is going to wreak you the moment you get back home." He pushed me toward his door with strength.

I held back my tears as I pulled up my breeches and pushed down my misery. I left his room and grabbed my spare overnight bag. I filled it with enough outfits and sanitary items for the two days I would be gone. I took out Annette's picture and gave her a kiss for luck then dropped her into my vent hole with my other treasures. I didn't want her with me to witness the foulness of a blood bonding even if it was with the benign Master Leo.

I left with Master Claus and Master Bladrick without quarrel. The men seemed irritated and subdued. I knew they also had heard the gossip spread by Gretta. What surprised me was how angered the idea that I found my lust with Master Leo seemed to have made them. I didn't realize they were also the jealous type.

I barely made it through their apartment door when they tag teamed me with their molestations. They spent what felt like hours demanding I find interest in their foul grabbing, despite that chastity device which pissed them off as badly as it did me.

When at last they tired of trying to force my erection, as if anyone could with the shit they were doing to me, they played the London Bridge with poor old Maximillian.

I was humiliated, sore beyond imagination and sick to my stomach. That bloody urine was not good. When at last the two of them were sated, I wanted to shower, vomit, brush my teeth, and blow my head off with a shotgun.

However, they wouldn't allow me to be out of their sights. I saw the illness taking hold of the flesh and grew anxious over it. I decided for distraction I would find out what the Elders knew about this ancient blood bonding business.

I asked Master Claus what the blood bond was all about, and he happily informed me that this was a recognized marriage between a Dominant and his submissive, that counted even if a collar was busted. He went further to tell me that all submissives were permitted four blood bondings at once and no more could be made unless one of spouses died. *Dis rule used to be, but it has since changed to only two, one of each gender. In those days there was no distinction and four was allowed in total.*

The Elder told me that all blood bondings were supposed to be agreed upon by both parties. If one or the other believed they had been forced into this arrangement, then they could have it annulled. Such a thing could be accomplished by contacting the Head of the Voting Council and demanding it be undone. The one contesting the bond only had twenty-four hours to do this after it was witnessed or the marriage was no longer disputable for life.

I groaned as I realized this was the real reason, Master Jonas had Peter sent to the dungeon and bonded Christian Axel during the act. He didn't want to give Der Hund nor our Master Peter the chance to dispute the bond as the rape it was. Drexel had reported the act valid, and it was recorded for all time despite it being completely illegal even by Haus law.

Master Bladrick then chimed in that a blood bonded mate had full authority to the special services of their submissive. He or she could deny this privilege to anyone they wished.

There was a sudden silence among the two Elders when Master Bladrick said that. I felt a chill run down my spine. For starters, no doubt Master Jonas was about to win his case. It was the law that he could block all the other Masters.

Secondly, I realized with much terror, my Elder Masters were now alerted to this scary fact. Thanks to my stupidly asking about it, they understood their weakness at once. I could almost see it in their eyes that they were plotting a blood bonding with me.

I sat there wishing I had kept my mouth shut and waited to see what Master Leo knew of this shit. Fuck, I really stuck my foot into the trap this time. The two Elder schwulers excused themselves to nap, but I was not fooled. They were sneaking off to make their plans to marry the Priceless Mad Maxx.

I had to get out to see Master Leo somehow. I knocked on Master Claus's bedroom door and asked if I could go down to the kitchen for dinner. It was late and I had not eaten. He was in there whispering with Master Bladrick. I knew I was so fucked this time.

I winced when he yelled out. "Ja, I will call down an order. You pick up our tray too, then get back up here and eat with speed. Bladrick and I will be requiring your services in an hour or so. Hurry up, Mad Maxx." I took off for the door not needing to hear that shit a second time.

I was most aware that the next London Bridge would see the mingled blood of my Elder Masters. If they were successful, it would only leave one single blood bonding slot left. The Elders would no doubt call Master Jonas to witness.

Once the Vampire realized the plot, Master Leo would be blocked. I had to hurry up and find him or all would be lost for Leo and me.

I rushed to Master Leo's apartment and banged on his door as if the Haus was on fire. He answered looking as panicked as I felt.

"Leo, I have no time to explain but you must blood bond with me right this second. If you deny me, then come tomorrow I can kiss our home together goodbye." I nearly screamed out in a ramble.

Master Leo grabbed his chest looking pale. "Oh, meine Gott. Jonas, he is going to try to block the Elders' leashes." I nodded most grateful he understood without my needing to explain my crazy sounding statement.

He took a deep breath. "Calm down Christian Axel. Now listen to me carefully. I need to know two things. ne, how long do we have to do this? Two, is there any other seed in you at this moment?"

I narrowed my eyes at those weird questions. "Uhm, we have a little more than an hour, and ja I just dealt with Master Claus and Master Bladrick. I haven't had time to cleanse yet. I need the blue water."

Master Leo sighed. "Shit, okay look, I could have you shower here, but I am sure they could question it. You must be free of any disputable fluids that are not of the one you bond to. Wait, I have an idea. Okay, you go to the pool this minute and get into the water. I will call Cora and tell her I wish to make peace. She will show up and find the two of us in the middle of the bonding. Then no one can dispute my claim because it was witnessed while in process. Ja, this is perfect. Now go and hurry. I will call the bitch. See you in fifteen minutes. Oh, and meine heart, stay out of the deep end." He kissed me quickly on my lips and I took off in a full run for the stairs to follow his commands.

I jumped the straggling silvers who were heading to their Masters for the night, and I avoided the black collars rushing about to finish their daily chores. I made it to the pool room in record time. I stopped at the entry and looked inside. I near screamed in delight to find it abandoned. I threw off my clothing and jumped into the deep in with my shirt flying off at the last second.

I went deep in the water from my dive. I used the skills Master Malfred taught me to come back to the surface. I swam across the pool feeling refreshed, relieved, and grateful that I had been able to avoid a huge disaster at the last second.

I splashed about giggling and enjoying my frolicking in the curative blue liquid with a song in my heart. I was about to blood bond to my Leo. What a joyful day to know that Master Jonas could never again block my lover from his rights to my services.

I saw Master Leo's silhouette slipping through the entry. I swam to the steps and got out ready to aid him in undressing. I rushed towards him with a huge smile on my face that immediately fell to a frown.

Master Malfred came walking briskly in my direction, his own smile building across his mouth. "Ah, Mad Maxx. What a pleasure to meet you here again. Wait a minute, are you alone again? Didn't I tell you it is dangerous for you to be running around here without any Dominant guardian? I see you didn't listen." He smiled even wider.

I shook my head. "Master Leo is on his way, Master. I am not alone. I followed his instructions to come and take a quick dip in the pool. I wonder if I may ask you to give him and I, a bit of privacy. Maybe you can hold off swimming for thirty minutes?"

He motioned me to kneel, which I did without quarrel hoping he would do me the favor I requested. "Well, you see I must deny your request, meine taube. If I leave this minute, you and Leo would blood bond. I cannot let that happen."

I gasped then looked up at Master Malfred in shock. "What? I don't understand what you are saying, Master."

Master Malfred began to unbutton his cuffs. "I know you don't, meine treasure. You see, I cannot let Leo take one of the spots I have already laid claim to. Now, I would like to say this won't hurt, but I am afraid it does, a great deal. Of course, this you already know with one blood bond already. I am here to take care of the other three. Grisham, Peter, Gustov, hurry this shit up. I have the boy. Bring that fucking Leo and keep him quiet. Cora will be here for the witnessing in thirty minutes. I want to get this over with." I trembled in terror as I saw the huge brute Grisham, Peter and Gustov come through the doors.

Peter was holding a kicking and fighting Leo. They had knocked him down and tied his wrists behind him. Gustov closed the huge doors and bolted them. I looked wildly around the room for an escape.

It was then I noticed Grisham had a bandage on his right ear. He was the second rapist. I shot a look of fear at Master Malfred who was now removing his suit jacket. I whimpered as the sunglasses fell to the floor from his waistcoat pocket.

I groaned out. "You are Casper. Grisham was the rapist in the woods and in Barnim's apartment. You watched him attack me. Then drugged me and helped me home. Why?" I felt the tears welling, unable to accept I had fallen into another trap of Peter's and heartbroken that Leo was likely involved again.

Master Malfred chuckled. "You are too precious for words. Of course, Grisham and Julius were the rapists, meine taube. They both work for me. I told them to attack you. To be fair, Peter made up that part of the plan. I believe you know him well, ja? Well actually you know Grisham well too. Only Gustov here is new to you, meine taube. This fine fellow took that dumb bastard Julius's place in our little plot. I admit I did my best to turn you on Leo for you to kill the sonofabitch, but you are one loyal soul. You never gave into my tactics no matter how I tried to convince you of Leo's guilt as your rapist. I am sure given more time I could have gotten you to end him for me. However, too late now. Oh well, I guess you can't have everything you want, ja? That damned brute Jonas has moved my plans along faster than I would have liked. We must proceed with phase two without all my plans in place as I had hoped for." Gustov and Grisham had moved in to surround me like two giant wolves.

Peter dragged Leo to one of the pool chairs and with some struggle made him sit down. He then sat on the wriggling man's lap to keep him from escaping. I felt the room spinning as I noticed all three of the huge Dominants were undressing but never taking their eyes off me shivering in the middle of them.

I sniffed hard trying to regain my mind from the numbing fear building within. "Why did you not end me when I killed Julius. You saw it, but let it happen." I was trying to find a way to distract Master Malfred. I was preparing to attempt to run.

He shot a glance at Peter and the railing Leo, then looked back at me. "Julius tried to double cross us. He showed you, his face. The dumbass decided to carry out the personal vendetta for that bitch Beatrice. He also planned to break the rule that you were not to be killed. Grisham or I would have had to get rid of him, but you, meine taube, did the dirty work for us. Thank you for it. I must admit it was an impressive thing to see. Such a vicious little beast. Grisham and Peter had told me that you are quite a fighter. I am the believer after seeing that scene. I had hoped to obtain my first taste of my Priceless collar with you willing, but again, Jonas insists on making this an ugly, painful, and rushed experience for you. For that I apologize, but what must be done will be done. Grisham and Gustov will restrain you. I will go first. This will be my forbidden silver after today." He smiled at me as he unbuttoned his fancy slacks.

Grisham and Gustov reached down to grab my arms, but I slipped under the big men in a sudden move. They yelled, demanding I halt just as I dove headfirst into the deep end of the pool. I did my best to make it to the bottom ready to take a lung full of the curative water and end my life. I heard the splashes of three breaking the surface.

I closed my eyes and willed myself to take a deep breath just as the first Dominant grabbed me from the bottom and raced me to the air. I struggled and kicked but Grisham and Malfred dragged me along to the side as Gustov got out. He snatched my leash and collar pulling me with brutality from the water. The other two got out and ran to aid his trying to restrain my wild coughing, kicking, punching, and biting of the late Abelard's Dominant.

I was no match for the big men. They dragged me back to their discarded clothing then forced me to my back. Grisham and Gustov restrained my arms, pinning them above my head while Malfred managed to get between my kicking legs. I wailed in terror as the Dominant pawed and kissed on my flesh, working himself into a lustful frenzy. I could hear Leo begging them to let me go in the distance. Peter kept threating to "shut him up for good."

I began to sob loudly when I saw that Malfred had obtained an erection. He told Grisham to toss him his penknife. In pure horror I watched the man cut into the head of his cock. Then without another moment hesitation nor lubrication he forced his bleeding member into me. He thrust with much brutality making sure to rip my tissues and mingle our fluids. I gasped, screamed, and cried openly while Master Malfred sodomized me without mercy.

The boy went near catatonic while enduring this cruel rape before Malfred sped up his mount. Within moments of his increased thrusting, he fell over me panting in ecstasy with a smile of thrill on his cold lips. I openly sobbed and shuddered, still trying to struggle from the tight hold of his men. He loudly announced he had reached his orgasm and blood bonded successfully for my lifetime.

"You are now meine man. This is my lifelong dream come true. At last, I finally can lay claim that I possess the finest of all the collars, the Priceless himself. Ah don't cry, meine taube. You have served your Master well and brought me to pure rapture. I have not been this pleasured since my Tamina was only a young fresh girl. You are indeed the forbidden silver of legend. I admit I am angered to have to share you with these unworthy brutes for even this one time. However, if I am to hold you for meine own for all time, this foulest deed must be done to completion. Grisham, your turn. I warn you. If you put one more Gott damned mark on my collar's flesh, I will drown you in that pool, you jackal. I am sick of your harsh handling of my precious treasure. When you finish this taste of his metal, I better never see you around my Priceless again. You listening to me?" He looked at the brutal Grisham with a stern expression.

Master Grisham nodded. "Well, aren't you just the lucky one Malfred. I thought we would share this treasure. I do understand that the plans have changed but I am still pissed off about it. I have enjoyed him a great deal and am in no hurry to end my thrills. You better not forget your promises to me if I am to give up my tasting of his forbidden metal after this. If I have your word, then come take my spot. This boy is a fighter and stronger than he looks. I will be needing that penknife of yours Malfred." Master Malfred nodded then reached out and picked up his knife.

"You have my word, jackal. Gustov you will go last. You better hurry your intercourse Grisham. We are running out of time." He uncoupled and Master Grisham and he traded spots so rapidly I couldn't even attempt to break free.

To my horror I watched the brute Master Grisham cut his readied cock in the same fashion as Master Malfred before him in haste. I wailed in misery as the brute spit on the ground just before he forced himself into me and began his third violent raping of the boy.

I realized with hopelessness this was not just a gang rape for the thrill of the sex. The three of them were blood bonding me to them. Almost the entire voting Council would have the rights to the Priceless anytime they wished.

I could never blood bond to anyone else, unless one or more of them, died.

To Be Continued in Book 6: Mastermind Malfred

Alexandria May Ausman began demonstrating severe psychotic episodes while still in her teens. She was abandoned by her family after being diagnosed with Paranoid Schizophrenia at age sixteen. Forced to struggle with this devastating illness alone, she has suffered medication resistant symptoms, numerous hospitalizations, homelessness, exploitation, and an uncaring mental health system.

Despite the hardships, Alexandria managed to raise two healthy children to adulthood and has four beautiful grandchildren. She obtained a bachelor's degree in psychology and held a job as a child abuse investigator. In 2003, she began a career as a diagnostic psychologist while working towards a Master's in psychology.

Alexandria never forgot the experience of 'slipping through the cracks.' She worked tirelessly to help people suffering severe mental illness and/or all types of abuse have access to necessary services for over seventeen years.

In 2017, she was published and became a professional model of "goth fashion.' and won the World Gothic Models contest in 2018. She holds the title of World Goth Queen for life.

Alexandria began writing several series of fictional novels after a catastrophic return of psychotic symptoms in 2019. She obtained the Killer Nashville Falchion Award as Best Southern Gothic writer in August 2023, and is a finalist for her book Delusion of the Collar and the Key.

Today, Alexandria is retired, and homebound due to crippling symptoms of Schizophrenia. She currently lives in Tallahassee, Florida, with her loving husband and a loyal support dog.